COSYSMO 2.0: Estimating Systems Engineering Reuse

Jared Fortune

COSYSMO 2.0: Estimating Systems Engineering Reuse

Constructive Systems Engineering Cost Model, Version 2.0

VDM Verlag Dr. Müller

Impressum/Imprint (nur für Deutschland/ only for Germany)

Bibliografische Information der Deutschen Nationalbibliothek: Die Deutsche Nationalbibliothek verzeichnet diese Publikation in der Deutschen Nationalbibliografie; detaillierte bibliografische Daten sind im Internet über http://dnb.d-nb.de abrufbar.

Alle in diesem Buch genannten Marken und Produktnamen unterliegen warenzeichen-, marken- oder patentrechtlichem Schutz bzw. sind Warenzeichen oder eingetragene Warenzeichen der jeweiligen Inhaber. Die Wiedergabe von Marken, Produktnamen, Gebrauchsnamen, Handelsnamen, Warenbezeichnungen u.s.w. in diesem Werk berechtigt auch ohne besondere Kennzeichnung nicht zu der Annahme, dass solche Namen im Sinne der Warenzeichen- und Markenschutzgesetzgebung als frei zu betrachten wären und daher von jedermann benutzt werden dürften.

Coverbild: www.ingimage.com

Verlag: VDM Verlag Dr. Müller Aktiengesellschaft & Co. KG
Dudweiler Landstr. 99, 66123 Saarbrücken, Deutschland
Telefon +49 681 9100-698, Telefax +49 681 9100-988
Email: info@vdm-verlag.de
Zugl.: Los Angeles, University of Southern California, Ph.D. Dissertation, 2009

Herstellung in Deutschland:
Schaltungsdienst Lange o.H.G., Berlin
Books on Demand GmbH, Norderstedt
Reha GmbH, Saarbrücken
Amazon Distribution GmbH, Leipzig
ISBN: 978-3-639-28616-8

Imprint (only for USA, GB)

Bibliographic information published by the Deutsche Nationalbibliothek: The Deutsche Nationalbibliothek lists this publication in the Deutsche Nationalbibliografie; detailed bibliographic data are available in the Internet at http://dnb.d-nb.de.

Any brand names and product names mentioned in this book are subject to trademark, brand or patent protection and are trademarks or registered trademarks of their respective holders. The use of brand names, product names, common names, trade names, product descriptions etc. even without a particular marking in this works is in no way to be construed to mean that such names may be regarded as unrestricted in respect of trademark and brand protection legislation and could thus be used by anyone.

Cover image: www.ingimage.com

Publisher: VDM Verlag Dr. Müller Aktiengesellschaft & Co. KG
Dudweiler Landstr. 99, 66123 Saarbrücken, Germany
Phone +49 681 9100-698, Fax +49 681 9100-988
Email: info@vdm-publishing.com

Printed in the U.S.A.
Printed in the U.K. by (see last page)
ISBN: 978-3-639-28616-8

Table of Contents

List of Tables

List of Figures

Dedication

This book is dedicated to my beautiful wife, Elizabeth.

Chapter 1: Introduction

1.1. Motivation for Estimating Systems Engineering Reuse

All systems today have some degree of legacy considerations. Industry is rapidly going away from building systems from scratch. Instead, system upgrade and development spirals dominate today's programs. In this environment, the concept of reuse becomes an ever increasing subject in systems engineering (Valerdi, Rieff et al. 2007).

Technology is increasingly systemic in nature and the development and production of complex systems requires developers to simultaneously invest in several complementary areas of capability (Garud and Kumaraswamy 1995). Due to resource constraints, such an investment is difficult to make for every system, especially complex systems. As system complexity increases or resource availability becomes limited, a common occurrence with aerospace systems, systems engineers are frequently asked to (or seek to) leverage previously developed systems engineering products to reduce cost, schedule, and/or risk. The concept of systems engineering reuse is the utilization of previously developed systems engineering products including requirements, test plans, and interfaces; however, the effects of such reuse have historically been difficult to quantify, and therefore justify. In theory, by reusing previously developed products, a systems engineer should be able to reduce the amount of systems engineering effort required to develop a new system.

Despite the importance of systems engineering activities to the successful development of a complex system, systems engineering cost estimation techniques have been relatively immature and until this research, no cost models existed to account for the effect of systems engineering reuse. Historically, the cost of systems engineering has been metric-based, estimated as a percentage of the cost of the total system, hardware, or software, or through rules of thumb. To provide quantifiable justification for the cost of systems engineering expected for a system of interest, the Constructive Systems Engineering Cost Model (COSYSMO) was developed in 2005 at the University of

1

Southern California Center for Systems and Software Engineering (USC-CSSE); the first general-use parametric systems engineering cost model (Valerdi 2005).

Since its publication, COSYSMO has been widely accepted in both industry and academia. To date, four proprietary versions of COSYSMO have been developed by aerospace contractors, multiple commercial cost estimation software packages have incorporated the tool, and graduate engineering courses at University of Southern California, Massachusetts Institute of Technology, Naval Postgraduate School, University of California San Diego, Southern Methodist University, and George Mason University have brought the model into their coursework on cost estimation.

Approximately one year after the publication of COSYSMO, practitioners noted that large errors were being observed between certain cases of model estimates and actuals. Upon further review, these variations were discovered to be attributable to programs that reused a significant number of previously developed systems engineering products (Valerdi, Gaffney et al. 2006). Reuse is not adequately handled by the initial version of COSYSMO due to an underlying assumption of the model that all aspects of system development and related systems engineering activities are new. One exception is the COSYSMO Tool Support cost driver, which adds tool development effort if reusable tools are available. In other words, COSYSMO assumes that all systems engineering activities and resulting products will need to be "built from scratch" with no previous systems engineering products being reused. While this assumption does not affect systems that do not leverage any previously developed systems engineering products, it does affect estimates of systems that do reuse. The effect of not accounting for reuse is the frequent, and potentially significant, over-estimation of the amount of systems engineering effort for a system under development. COSYSMO provides reasonable and valuable estimates of total systems engineering effort for systems, but does not have the ability to estimate systems that reuse systems engineering products. Although the concept of reuse estimation was explored in previous COSYSMO reserach,

2

the lack of supporting data on systems engineering reuse in the original COSYSMO calibration data set left the concept untested, not validated, and ultimately unincorporated in the published model (Valerdi 2005).

At COSYSMO Working Group meetings the following year, accounting for systems engineering reuse was identified as the most critical area of research on COSYSMO by representatives from over eight industry and academic organizations (Rieff et al. 2007). Given the success of COSYSMO from both an industry and academic perspective, further development and improvement of the model was both needed and justified (Valerdi et al. 2008). This leads to the principal question addressed by this research:

What technical and organizational characteristics drive the amount of reduced or added systems engineering effort due to systems engineering product reuse?

The model presented in this book, COSYSMO 2.0, as well as other contributions to the field of systems engineering produced during the course of this research, which facilitated the development of the model and support its application, provide systems engineers with the means to answer this question.

1.2. Fundamentals of Systems Engineering Reuse

Nearly every engineering discipline has a definition for reuse, all with a common theme of applying an existing solution to a new application. Examples include:

Industrial Engineering: *To assemble a product from existing components and limit the creation of new components to ones that do not exist (Robertson 1996).*

Electrical Engineering: *The use of an asset in the solution of different problems (IEEE 1999).*

Software Engineering: *The ability to apply a solution to specific instances in other domains (Prieto-Diaz 1993).*

By any definition of the term, reuse is not a new concept. Early forms of reuse include the repetition of mathematical models and algorithms across problems to ensure correct calculations (Prieto-Diaz 1993). The construction and automobile industries rely

3

heavily on the reuse of key components and parts (de Judicibus 1996). Even the utilization of engineering specifications and standards, essential to any engineering practitioner, is a form of reuse.

Fundamentally, reuse is the result of a natural human problem solving technique whereby people determine if a problem they are faced with has already been solved, if they have an existing solution to a similar problem that can be adapted, or if the problem is unprecedented and needs to be decomposed into a smaller set of sub-problems (Mili et al. 2002). Reuse is different from the concept of re-engineering in that re-engineering occurs when an existing system is transformed into another system, whereas reuse occurs when a product is re-applied to a new application (Lam and Loomes 1998).

This research is focused on systems engineering reuse and specifically, how reuse impacts the expected amount of systems engineering effort for a system. For systems engineering, more applicable definitions of reuse include:

The repeated use of an application in different places of the design of parts, manufacturing tools and processes, analysis, and particularly knowledge gained from experience; using the same object in different systems or at different times in the same system (Allen et al. 2001).

The use of systems artifacts and processes in the development of solutions to similar problems (Whittle et al. 1996).

Unlike most other engineering disciplines, which are generally hardware or software product-focused, systems engineering discipline is mainly activity-focused and its products are generally limited to analysis and supporting documentation. Therefore, systems engineering reuse is defined as the utilization of previously developed systems engineering products, such as requirement documents or test plans. By reusing a systems engineering product, some amount of the activities associated with generating the product may not need to be repeated, thereby reducing the expected amount of systems engineering effort for the system.

1.3. Proposition and Hypotheses

4

With the motivation for this research identified and the definition of systems engineering reuse established, the central proposition and hypotheses are proposed. The central proposition of this research is:

There exists a subset of systems engineering activities from a defined application domain for which it is possible to create a parametric model that will estimate the amount of systems engineering effort throughout specific life cycle phases (a) for a specific system of interest that involves the reuse of systems engineering products, and (b) with improved statistical accuracy over COSYSMO.

The statement provides the underlying goal of COSYSMO 2.0 by clarifying its solution space. The *subset of systems engineering projects* provides a homogenous group of projects from which the model can be based, characterized by discriminators such as domain, organization type, and productivity. The *defined application domain* specifies a domain from which supporting data was obtained and the model can be based. The term *parametric* implies that a given equation represents a function that is characteristic of Cost Estimating Relationships in systems engineering. The identification of *specific life cycle phases* establishes the phases of the life cycle of a system in which systems engineering activities primarily occur. The *reuse of systems engineering products* implies the reuse of systems engineering products that are estimated by COSYSMO size drivers and COSYSMO 2.0 reuse categories. The term *better statistical accuracy* implies COSYSMO 2.0 will have greater estimation power than COSYSMO, which in most cases is capable of estimating systems engineering effort within 30% of actuals, 50% of the time, or PRED(.30) = 50%.

The statement in part (*a*) is meant to define a specific level in the system hierarchy in which the model can be used. For example, the system of interest for a prime contractor may be the entire system; whereas the system of interest for a subcontractor may be a subsystem element. The statement in part (*b*) is the relevant measure that can be used as criteria to determine the model's ability to estimate systems engineering effort.

5

This relevant measure is similar in form to the other models in the Constructive Cost Model (COCOMO) suite and will be improved as more data is collected.

The central proposition was validated through the use of the scientific method (Isaac and Michael 1997). In terms of scientific inquiry, the model was validated through one quantitative and two qualitative hypotheses:

H1: The requirements size driver can be further decomposed into New, Designed for Reuse, Modified, Adopted, Managed, and Deleted categories of reuse, each with corresponding definitions and weights, and function as good predictors of equivalent size from a systems engineering standpoint.

Multiple iterations of these categories were applied to the model and their results were examined until a version was obtained where the first hypothesis could be properly evaluated.

H2: By comparing the normative and descriptive models for managing systems engineering reuse, the systems engineering reuse framework provides a prescriptive model for enabling organizations to take advantage of reuse in systems engineering.

Multiple iterations of the framework were created through interviews with subject matter experts until a version was obtained where the second hypothesis could be assessed through mini-case studies.

H3: The COSYSMO 2.0 model and the associated systems engineering reuse framework help organizations think differently about systems engineering reuse, and its associated costs and benefits.

The third hypothesis was developed to test the impact of the model on systems engineering organizations. This hypothesis was validated through interviews with subject matter experts.

Chapter 2: Background and Related Research

2.1. State of the Art

The reuse model presented in this book is derived from and extends the original COSYSMO tool. Before COSYSMO 2.0 can be proposed, the context, structure, and scope of COSYSMO needs to be described.

First, COSYSMO functions in the domain of systems engineering. For context, some definitions of systems engineering include:

> The application of scientific and engineering efforts to 1) transform an operational need into a description of system performance 2) integrate technical parameters and assure compatibility of all physical, functional and program interfaces in a manner that optimizes [or balances] the total system definition and design and 3) integrate performance, producibility, reliability, maintainability, human factors, supportability and other specialties into the total engineering effort (Blanchard and Fabrycky 1998).

> The application of a general set of guidelines and methods useful for assisting clients in the resolution of issues and problems which are often large scale and scope. Three fundamental steps may be distinguished: a) problem or issue formulation, b) problem or issue analysis and c) interpretation of analysis results (Sage 1992).

> An interdisciplinary approach and means to enable the realization of successful systems. It focuses on defining customer needs and required functionality early in the development cycle, documenting requirements, and then proceeding with design synthesis and system validation while considering the complete problem. Systems engineering considers both the business and the technical needs of all customers with the goal of providing a quality product that meets the user needs (INCOSE 2007).

Second, COSYSMO is structured as a parametric cost model. A parametric model is based on relationships between technical, programmatic, and cost characteristics of a particular system (ISPA 2007). As such, COSYSMO uses values from 18 systems engineering drivers, identified in Table 1, to account for product, platform, personnel, and project factors associated with the systems engineering effort (Valerdi 2005).

7

Size Drivers (4)	Cost Drivers (14)
Number of System Requirements	Requirements Understanding
Number of Major Interfaces	Architecture Understanding
Number of Critical Algorithms	Level of Service Requirements
Number of Operational Scenarios	Migration Complexity
	Technology Risk
	Documentation
	Number and Diversity of Installations
	Number of Recursive Levels
	Stakeholder Team Cohesion
	Personnel/Team Capability
	Personnel Experience/Continuity
	Process Capability
	Multisite Coordination
	Tool Support

Table 1 COSYSMO Size and Cost Drivers

The values for each of these drivers are inputted into the COSYSMO operational equation, presented in Equation 1 (Valerdi 2005).

Equation 1 $\quad PM_{NS} = A \cdot \left(\sum_{k} (w_{e,k}\Phi_{e,k} + w_{n,k}\Phi_{n,k} + w_{d,k}\Phi_{d,k}) \right)^{E} \cdot \prod_{j=1}^{14} EM_{j}$

Where:

PM_{NS} = effort in Person Months (Nominal Schedule)

A = calibration constant derived from historical project data

k = {Requirements, Interfaces, Algorithms, Scenarios}

w_x = weight for "Easy", "Nominal", or "Difficult" size driver

Φ_x = quantity of "k" size driver

E = represents (dis)economies of scale

EM = effort multiplier for the j_{th} cost driver.

Third, the scope of COSYSMO is based on international standards for both the life cycle phases of a system and the systems engineering activities. COSYSMO estimates the effort associated with the systems engineering activities during the first four phases of

8

the system life cycle, as defined by ISO/IEC 15288 System Life Cycle Processes, depicted in Figure 1 (ISO/IEC 2002).

Figure 1 ISO/IEC 15288 System Life Cycle Phases

The systems engineering activities within these life cycle phases are identified by ANSI/EIA 632 Processes for Engineering a System, which helps to define a typical work breakdown structure that can be used for estimating cost (ANSI/EIA 1999). The thirty-three systems engineering activities defined within ANSI/EIA 632 are illustrated in Appendix A.

Given that COSYSMO has its foundations in systems engineering standards, one of the first strategies for developing the reuse extension was to examine various systems engineering standards, handbooks, and key references to understand how systems engineering reuse was addressed in the literature.

Maier and Rechtin's *The Art of Systems Architecting* presents dozens of systems architecting heuristics, which document experiential knowledge to pass to less experienced architects and engineers (2002). Several heuristics were observed to focus on topics and issues related to reuse, but none that addressed reuse specifically. Some reuse-related heuristics and their applicability to reuse are listed in Table 2.

Reuse-Related Heuristic	Applicability to Reuse
If you don't understand the existing system, you can't be sure you're architecting a better one.	Need to understand the product being reused
The test of a good architecture is that it will last. The sound architecture is an enduring pattern.	Well designed products are reusable
In architecting a new system, by the time of the first design review, performance, cost, and schedule have been predetermined.	Reuse needs to be planned
Good products are not enough. Implementations matter.	Domain application is a key factor in reuse success

Table 2 Systems Engineering Reuse Heuristics

Although none of the heuristics published by Maier and Rechtin explicitly mention reuse, they are relevant and helped to guide the development of the revised model.

Blanchard and Fabrycky's *Systems Engineering and Analysis* explains a variety of systems engineering process models and identifies key systems engineering activities, but no mention of reuse was observed (Blanchard and Fabrycky 1998). With the concept of reuse not addressed in either of two seminal systems engineering texts, the review moved to examining systems engineering documents from the industrial practice.

Table 3 highlights the results of an examination of select systems engineering handbooks and processes. All of the documents reviewed at least mentioned the term reuse, but the concept of reuse was hardware or software product-focused and not systems engineering product-focused.

				Discuss Reuse?	Product Reuse?	Systems Engineering Reuse?
Systems Engineering	Handbooks	INCOSE	Systems Engineering Handbook (v. 3.1)	✓	✓	✓
		NASA	Systems Engineering Handbook (2004)	✓		
		NASA	Systems Engineering Processes and Requirements (7123.1A 2007)	✓		
		IEEE	Systems Engineering (1220-2005)	✓		
		EIA	Processes for Engineering a System (2003)	✓		
Software Engineering	Standards & Procedures	IEEE	Software Life Cycle Processes (1517-1999)	✓	✓	

Table 3 Systems and Software Engineering Document Review Results

In the INCOSE Systems Engineering Handbook (INCOSE 2007), one instance of reuse was mentioned, the reuse of a Systems Engineering Management Plan template that could be tailored to the characteristics of individual projects. While this was an example of reuse of a systems engineering product, no other details on reuse were presented. Other than this example, no other systems engineering documents reviewed had any mention of non-hardware or software reuse. However, during the review, it was observed that some of these systems engineering documents made reference to software engineering documents. Knowing the close connection between the systems and software engineering domains, software engineering documents were reviewed for additional insight on reuse (Maier 2006).

An initial review of the software literature revealed a large number of documents that addressed reuse. At this point in the research, it was realized that a more exhaustive literature review on reuse in the software engineering domain would be required. The rest of this section summarizes literature on reuse in the software engineering domain and presents several observations about reuse from the software engineering literature review, which are applicable to systems engineering reuse and helped to guide the development of COSYSMO 2.0 (Fortune and Valerdi 2008).

IEEE defines software reusability as the degree to which an asset can be used in more than one software system, or in the building of other assets (IEEE 1999). Similarly, Prieto-Diaz states that software reuse is the ability to apply a solution to specific instances in other domains (1993). The idea of reusing software was first discussed publicly in 1969 by Bell Laboratories, when Bell proposed to make software development more "industrialized" instead of "craft-based" (Isoda 1996). Over the past few decades, software reuse has been described as a means for enabling projects to achieve higher quality, increased productivity, shorter development schedules, reduced overruns, and improved leveraging of technical skills and knowledge (Boehm 1981) (Boehm and Scherlis 1992) (Basili et al. 1987) (Lim 1998). To date, dozens of models have been developed to estimate a wide range of parameters associated with software reuse; more extensive summaries of software reuse metrics and models can be found in (Frakes and Fox 1995, Lim 1996, Poulin 1997, and Wiles 1999).

The motivation most often stated in the literature for software reuse is a reduction in the cost (and, increasingly, calendar time) of developing new products by avoiding redevelopment of capabilities and increasing productivity by incorporating components whose reliability has already been established (Bollinger and Pfleeger 1992) (Poulin and Caruso 1993). For example, the reuse of software code can result in fewer program faults and repeat mistakes can be avoided (Antelme et al. 2000) (Selby 2005).

Observation #1: Reuse is done for the purpose of economic benefit, intending to shorten schedule, reduce cost, and/or increase performance.

Naturally, once an organization finds a product that performs, they want to replicate its success. When successful, a software reuse program can result in cost savings between 10-35% (Stephens 2004); however, reuse should not be considered a "silver bullet". Organizations frequently predict that reuse will result in huge increases in productivity (Tracz 1988) or overstate their capabilities and overestimate the chances for reuse success (Szulanski and Winter 2002). Even small modifications of a product can negate any potential reuse benefits, therefore making it more efficient to start with a new product than a reused one (Glass 1999). On the other hand, a well-planned and executed reuse program also reduces life cycle maintenance costs due to fewer components requiring separate maintenance (Boehm et al. 2004).

Observation #2: Reuse is not free, upfront investment is required.

Reusable products are hardware, software, process, or knowledge focused (Basili et al. 1987). These products can be requirements, designs, code, tests, test cases, architectures, documentation, interfaces, and plans (Whittle et al. 1996) (Lim 1998). Reusable products can also include certification processes, configuration management records, quality records, and verification data (Lougee 2004); even products such as budgets, SWOT analyses, contracts, and prototypes are reusable (Cybulski et al. 1998). In addition to the wide variety of reusable products, the processes captured within and associated with the development of products are also essential for successful reuse. Boehm states that software reuse itself needs to be process orientated (1999). Reuse processes should be formal and institutionalized to capture reuse principles, produce quality results, and be repeatable (Mili et al. 2002) (Whittle et al. 1996).

Observation #3: Hardware, software, processes, and knowledge are all reusable products.

According to Tracz, software reuse is not something that will just happen (Tracz 1988). For reuse to be successful, it must be planned from the onset of the project, as the difficulty of implementing reuse becomes increasingly harder as system development progresses (Finkelstein 1988). Because of this, reuse is most frequently successful when it is planned (i.e. strategic or systematic), compared to an opportunistic (i.e. non-planned

13

or ad hoc) approach (Dusink and van Katwijk 1995). Opportunistic reuse is the idea that a development can be stopped at selected life cycle phases, potential reusable products can be reviewed for applicability, and reuse of those components can occur (Prieto-Diaz 1996). Opportunistic reuse is characterized by unplanned, short-term solutions (Lim 1998), although simply putting components or products together is usually unsuccessful and frequently results in negative impacts to project schedule and total effort (Garlan et al. 1995). Planned reuse is characterized by a predetermined and well-coordinated process (Lim 1998) (IEEE 1999). Planned reuse is similar to product line engineering (Beckert 2000), which is the strategic and planned use of architectures and components across development efforts (Wiles 1999). The success of planned reuse over opportunistic reuse can be attributed to the fact that a planned approach helps an engineer to assess the impact of reuse on the system beforehand and prepare for the potential issues (Lam et al. 1997).

Observation #4: Reuse needs to be planned from the conceptualization phase of system development.

One of the most commonly discussed aspects of reuse in the software engineering literature was that reuse is not a technical problem, it is a psychological, sociological, or economic one (Tracz 1988). From an economic perspective, a viable business case for reuse needs to exist before such a strategy should be pursued (Reifer 1997). A business case should not focus purely on the potential economic benefits expected from reusing software or explain the quality of the product being reused, but rather the capabilities of a skilled workforce and knowledge of the system that the products were derived from (de Judicibus 1996). From a sociological standpoint, products are rarely built from scratch, personnel do not typically forget years of training and experience, and this knowledge usually exists somewhere in an engineering or corporate memory (Whittle et al. 1996). Subsequently, knowledge and personal experience, can be reused (Basili and Romach 1991); however, capturing such information is often challenging (Moore 2001).

14

In a brief literature review of the knowledge management domain, it was observed that capturing and transferring knowledge can be laborious, time consuming, and difficult; not costless and instantaneous as commonly perceived (Szulanski and Winter 2002). Knowledge reuse can be having experienced personnel tell others where to find related information, how to solve a particular problem (Malhotra and Majchrak 2004), or how to apply other project specific information (Cooper et al. 2002). However, as mentioned previously, capturing knowledge and making it available for reuse is a major challenge. Reusable knowledge often exists only within a person (Szulanski and Winter 2002) and an organization needs to have adequate processes in place to capture, store, recall, and apply that information.

Observation #5: Reuse is as much of an organizational issue as it is a technical one.

A major evaluation criterion for reuse is domain compatibility (Konito et al. 1996). Successful reuse requires an understanding of the technology domain of interest in order to recognize what should be reused and how to accomplish it successfully (Tracz 1995). Analyzing a domain is: 1) the process of identifying, collecting, organizing, and representing the relevant information and 2) based upon the study of existing systems and their development histories (Mili et al. 2002). The failure to perform systematic and rigorous domain analysis accounts for the failure of many reuse programs (Anthes 1993) (Lam et al. 1997), as the potential for reuse cannot be judged by only looking at inputs and outputs of a system (Wymore and Bahill 2000). In addition, Selby determined that even for reuse within a related domain, the benefits of reuse do not scale in a linear fashion (Selby 2005). As project size or complexity increases, the application of reuse cannot be expected to deliver benefits along linear trend.

Observation #6: The benefits of reuse are limited to related domains and do not scale linearly.

The observations, summarized in Table 4, made during this review of the software engineering literature led to a survey on how systems engineering practitioners address reuse. The results of this survey are presented in the following chapter.

	Observation
#1	Reuse is done for the purpose of economic benefit, intending to shorten schedule, reduce cost, and/or increase performance.
#2	Reuse is not free, upfront investment is required.
#3	Hardware, software, processes, and knowledge are all reusable products.
#4	Reuse needs to be planned from the conceptualization phase of system development.
#5	Reuse is as much of an organizational issue as it is a technical one.
#6	The benefits of reuse are limited to related domains and do not scale linearly.

Table 4 Systems and Software Engineering Observations from the Literature

2.2. State of the Practice

After conducting a review of the systems and software engineering literature on reuse, several modeling considerations and conceptual approaches were identified. With a more complete understanding of the theoretical aspects of reuse, a better grasp of the practical approaches to reuse was necessary. To obtain this, a survey was created and distributed to representatives of various systems engineering organizations who had previously supported COSYSMO development efforts. The results of this survey yielded very valuable research results and helped to further guide the development of the reuse model.

A literature review of how the systems and software engineering domains handle reuse led to the identification of several key considerations for successful reuse. While this exercise was informative at providing the background for the reuse subject, it raised additional questions about how systems engineering reuse could be captured, quantitatively or qualitatively. Furthermore, industry perspectives and practices on reuse were still unknown. To obtain an industry perspective on the subject, a survey was developed and distributed to USC-CSSE affiliates and other interested parties (Fortune et al. 2008). The goal of the survey was to obtain more focused answers on systems engineering reuse that would facilitate the development of a COSYSMO reuse extension. Using the observations obtained from the literature review, some key questions on the systems engineering reuse were formulated, identified in Table 5. In addition to these questions, the survey solicited opinions on the issues of how reuse scales with system complexity.

| How do systems engineering organizations define reuse? |
| What systems engineering products are typically reused? |
| When in the system life cycle does reuse occur? |
| What contributes to successful or unsuccessful reuse? |

Table 5 COSYSMO 2.0 Reuse Survey Key Questions

From these questions, a ten-question survey (Wang, Valerdi and Fortune 2008) was developed, presented in Appendix B. The survey was distributed to subject matter experts who either previously supported the development of the COSYSMO tool and/or were involved with systems engineering organizations. Eleven responses were received from six different aerospace and engineering contractors: BAE Systems; General Dynamics; Lockheed Martin; Orbital Sciences; Raytheon; and Reynolds, Smith, and Hills.

The outcomes of the survey, presented in Figure 2 through Figure 6, can be distilled into three main results for how practitioners address and manage systems engineering reuse. These results are summarized in Table 6 and elaborated on in this chapter.

	Result
#1	Requirements reuse is only performed occasionally, but has the largest "benefit" associated with it.
#2	Reuse occurs more frequently early in the life cycle than later.
#3	Cost savings is the most promoted benefit for reuse.

Table 6 Reuse Survey Results

Requirements reuse was identified as being performed occasionally; less frequent than the reuse of any other the products mentioned in the survey, illustrated in Figure 2.

17

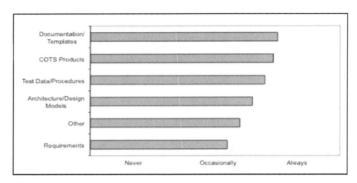

Figure 2 Frequency of Reuse of Systems Engineering Products

Result #1: Requirements reuse is only performed occasionally, but has the largest "benefit" associated with it.

This observation came from the applicability of requirements being reused relative to other products. For example, documentation is a product that is usually general enough to be applicable to multiple system development efforts. One subject matter expert stated that hundreds of systems engineering documents are available for potential reuse, but they are often more of a convenience, such as a document template, than a means of achieving a significant reduction in systems engineering effort.

If a requirement can be reused, most of the systems engineering products below the requirements level can be reused as well. Requirements are not too specific such that they cannot be applied to related systems and are not too generic such that they would fail to provide the capabilities to result in a sizable reduction in systems engineering effort, illustrated in Figure 3. However, a major risk area involves reuse across different nonfunctional requirement levels, as feasible architecture solutions are after discontinuous functions of these levels (Boehm et al. 2000). Additionally, even minor modifications to existing requirements can have significant, negative impacts to the potential benefits of reuse. In one example cited by Boehm, changing the requirement for a system to have a one second response time instead of a four second response time increased the total cost of the system by over 300% (Selby 2007).

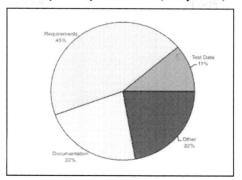

Figure 3 Products Cited as Most Effective at Providing a Benefit When Reused

Therefore, the challenge appears to be for a systems engineering organization to reuse applicable requirements without substantial modification.

The survey results indicate that systems engineering products are reused on a more frequent basis during the Conceptualization, Development, and Test and Evaluation life cycle phases, presented in Figure 4.

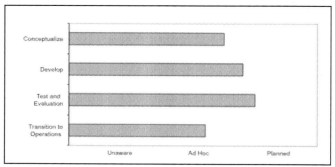

Figure 4 Extent of Reuse of Systems Engineering Products by Life Cycle Phase

Result #2: Reuse occurs more frequently early in the life cycle than later.

The majority of the systems engineering effort occurs in these phases and by the time the system reaches Transition to Operations, the development of the system is all but complete and limited opportunities for reuse are available. Additional discussion with a subject matter expert who responded to the survey stated that systems engineering reuse must be planned from the conceptualization phase because as the schedule progresses, it becomes more difficult to identify opportunities for reuse and reuse could have potentially unforeseen negative consequences. For example, reuse later in the life cycle will force a systems engineer to re-validate and re-verify all the interfaces that could potentially be affected by the reuse. One survey responder even went as far as saying that systems engineering reuse only occurs in the Conceptualization and Test and Evaluation phases.

Not surprisingly, cost savings was identified as the most promoted benefit for systems engineering reuse, illustrated in Figure 5.

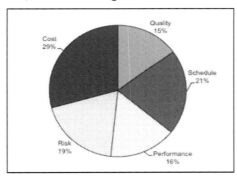

Figure 5 Promoted Benefits for Systems Engineering Reuse

Result #3: Cost savings is the most promoted benefit for reuse.

Although the realization of cost savings is difficult, the motivation for reuse appears to be the opportunity to reduce the amount of resources required. However, the other four benefits identified in the survey were equal in response rate and were not listed with significantly less frequently than cost. It was unclear if responders inherently associated risk, performance, schedule, or quality benefits as a means of achieving cost savings or if these factors are equally promoted. Despite schedule being ranked as the second most promoted benefit for reuse, additional discussions with survey responders cited risk reduction as the other major benefit. Reusing a product with a established heritage dramatically reduces the risk associated with a new system.

Survey responders were also asked about their opinion on how reuse benefits scale, increases or decreases, with system complexity. Most responders, citing an increasing number of system interfaces, believe that reuse benefits decrease non-linearly with system complexity, illustrated in Figure 6.

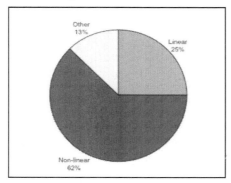

Figure 6 Scaling of Expected Reuse Benefits

Along with these quantitative answers, a number of descriptive answers about reuse were desired. The survey asked responders to identify reasons for reuse successes and failures, and for how reuse benefits scale with system complexity. The answers to these questions, along with the follow-up interviews with the responders, yielded extremely valuable insights into systems engineering reuse.

Based on the responses to the survey and the follow-up interviews, the most significant reason for the successful reuse of systems engineering products is the utilization of personnel with experience on previous system that developed the product. The identification of personnel as a key factor to the success of systems engineering

21

reuse supports Observation #5 from the software engineering literature review. Successful systems engineering reuse requires more than just reusable or proven products, non-technical factors such as personnel knowledge has a contributing role. Therefore, a strategy that only accounts for systems engineering reuse through a purely technical viewpoint would be incomplete. Successful systems engineering reuse is also attributable to the utilization of products with minimal to no modification. One subject matter expert stated that any modification of a product that exceeds approximately 20% will nullify any potential reuse benefit.

Comparatively, the most significant reason for the failure of systems engineering reuse is a product lacks a specific core capability. Even if a product is designed for reuse, it may be too generic to used for a specific application or from an unrelated domain, which supports Observation #6. For example, there could be a domain incompatibility, required modifications, or the product delivers multiple capabilities satisfactorily, but no single capability particularly well.

Chapter 3: Systems Engineering Reuse Framework

3.1. Motivation for Framework

While reviewing how organizations address reuse and developing the reuse cost model, the need for placing the issue of reuse in a larger perspective became evident. The reuse survey, COSYSMO Workshop discussions, and interviews with subject matter experts revealed that a mature process for reuse did not exist within or across most systems engineering organizations. The systems engineering field faces challenges that are both technical and social in nature, and systems engineering reuse is no exception (Valerdi and Davidz 2009). Without a standardized process for evaluating, executing, and documenting reuse, a cost model alone would be inadequate at improving the estimation of reuse. In addition to developing COSYSMO 2.0, a more holistic reuse framework was established to provide systems engineering organizations with an industry-validated process for handling reuse, additional considerations for successful reuse, and operational guidelines for the utilization of COSYSMO 2.0.

3.2. Framework Development

A framework provides a means of thinking about, classifying, or managing concepts (Garud and Kumaraswamy 1995) (Khadilkar and Stauffer 1996). Frameworks have been developed on wide range of concepts including software measurement (Kitchenham et al. 1995), marketing models (Larreche and Montgomery 1977), manufacturing techniques (Young 1992), and clinical theories (Fischer 1971). Process-related frameworks, such as the reuse framework presented in this book, illustrate relationships between elements in a particular process (Carlile 2004). A framework is useful when it addresses an abstract concept and provides perspective or context. From a practitioner standpoint, a framework can enable concept standardization, facilitate communication, and encourage knowledge sharing. From a research standpoint, a framework can support proposition formulation (Budros 1999), validate research methodology, and guide model development (Larreche and Montgomery 1977).

Frameworks are developed from observations from literature, discussions with subject matter experts, examples from industry, or a combination of methods (Young 1992). Validation of the framework occurs through iterations with subject matter experts and/or comparisons with industrial examples.

The framework presented in this book provides a prescriptive model for how an organization can address, execute, and manage the reuse of systems engineering products. A prescriptive model helps an organization transition from their current, or descriptive model, to a more normative, or ideal model (Valerdi, Ross, and Rhodes 2007). As was observed from the survey, workshop discussions, and expert interviews, a formal or even semi-documented process for reusing systems engineering products does not exist in most organizations. Therefore, the descriptive of model of the reuse process for most systems engineering organizations is informal and unplanned. Due to this lack of a well-defined process for reuse, which could lead to the inconsistent application of COSYSMO 2.0 and erroneous estimates, the reuse framework was developed to document a standard process and provide organizations with a means of transitioning towards a more ideal, planned, reuse process.

The criteria for a good framework are similar to the criteria for a good model (Valerdi 2008) (Boehm 1981). Table 7 lists the ten criteria for a good model and identifies how the reuse framework addresses each.

Model Criteria	Addressed by Reuse Framework
Definition	Identified categories of reuse, process boundaries
Fidelity	Included observations from literature and industry, as well as expert interviews
Scope	Applied to different types of systems engineering organizations
Objectivity	Generalizable to the aerospace and defense industries
Constructiveness	Captured a defined, standardized process
Detail	Illustrated a multi-level processes
Stability	Applicable Across Multiple Organizations
Ease of use	Provided prescriptive guidance to model
Prospectiveness	Addressed same life cycle phases as COSYSMO
Parsimony	Incorporated inputs from industry, validated with expert opinion

Table 7 Framework Criteria

3.3. Framework Description

The reuse framework, presented in Figure 7, is a multi-level figure identifying the actions that comprise the process of reusing systems engineering products. At its highest level, the framework identifies three major steps in the reuse process, identified by the large boxes: 1) Decide on Reuse, 2) Execute on Reuse, and 3) Archive Reuse. Each major step is made up of multiple sub-steps, identified by the small boxes, and support activities in various phases of the development life cycle, illustrated in the framework as Conceptualization, Development, and Evaluation. Conceptualization encompasses architecture evaluation, Development includes construction, and Evaluation covers verification and validation.

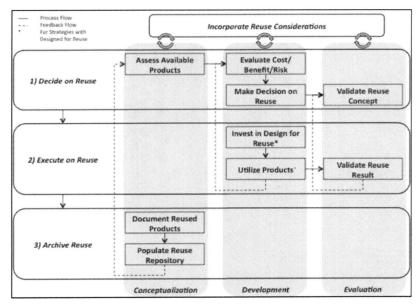

Figure 7 Systems Engineering Reuse Framework

The Decide on Reuse step involves examining the product to be reused and making a decision to reuse the product or not. In the Assess Available Products sub-step, an organization determines the products available for reuse. Products available for reuse include those designed to be reused as well as those not designed for reuse. Products designed for reuse, elaborated on in Chapter 4.3, were developed with reuse in mind and as a result are typically applicable to more systems, can provide greater benefit when reused, and/or will introduce less risk into the system when utilized. With a planned reuse strategy, the Assess Available Products sub-step would support activities early in the development cycle and may have an impact on the overall development effort. Conversely with an opportunistic reuse strategy, the Assess Available Products sub-step could occur at any point during the development of the system. In the Evaluate Cost/Benefit/Risk sub-step, the product identified as available for reuse is evaluated for how much effort will be required to reuse it, how much effort will be saved as a result, and how the reuse of this product will influence system risk. COSYSMO 2.0 would be applied during this sub-step to evaluate the amount of effort (cost or savings) associated with reusing a systems engineering product. In the Make Decision on Reuse sub-step, the evaluation of cost/benefit/risk of reuse is weighed and if reuse appears to be a favorable decision, the product should be reused; both the Evaluate and Make sub-steps support activities during the Development phase. In the Validate Reuse Concept sub-step, a confirmation is made that the expected benefits of reuse are documented and supported, and the decision to reuse the product is justified. This sub-step supports activities in the evaluation phase and enables the expected benefits to be compared with the actual benefits after reuse is completed, which can help inform future reuse decisions on the product.

The Execute Reuse step involves investing in and utilizing products for reuse, and validating the result of reuse. In the Utilize Products sub-step, products selected for reuse in the Decide for Reuse step or identified to be made reusable are executed upon and incorporated in the system development. Prior to this sub-step, for instances where a product is being designed for reuse, the Invest in Designed for Reuse sub-step exists. This investment sub-step occurs only when an organization intends to create a reusable product and therefore must invest in its development upfront. Both of these sub-steps support activities in the Development phase. In the Validate Reuse Result sub-step, which supports Evaluation activities, the actual benefit of reusing the product is compared against the expected benefit as validation that the reuse effort was worthwhile. In both the utilization and validation sub-steps, feedback loops back to sub-steps in the Decide on Reuse step exist. For the Utilize to Evaluate loop, once the product is utilized, it will influence the development of the system and affect the current and future evaluation of cost/benefit/risk. For the Validate Reuse to Validate Concept loop, after the effort has been completed, the actual reuse result will be compared with the expected reuse result.

Lastly, the Archive Reuse step involves capturing the product for future reuse opportunities. In the Document Reuse Product sub-step, the attributes, details, metadata, and product itself are recorded. Along with recording the reuse of the product, the information documented should be archived in a data repository for reuse. A reuse repository (Department of Defense 2004) not only stores the appropriate information about a product for future reuse, but it enables organizations to more effectively search and locate previously developed products that could potentially be reuse. These sub-steps support activities in the Conceptualization phase of future systems. As a result, a feedback loop exists between the reuse repository sub-step and the assess products sub-step to illustrate how information from the repository can influence the assessment of products that are available to be reused.

Throughout the entire reuse process, other factors that affect the successful reuse of products should be considered, identified by the box titled Incorporate Reuse Considerations at the top of the framework. These other factors, or intangibles, are not purely technical aspects to consider, but include factors such as personnel experience, familiarity with the product and heritage system, organizational structure (i.e. hierarchical or matrix), and existing documentation. The state of these considerations for a specific effort will influence the successful reuse of products. At every step in the reuse process, these considerations should be reviewed and their assessment may lead to a change in the reuse plan. For example, an organization may determine that products are available for reuse, but during the evaluation of the associated risks, the organization may conclude that no one is familiar with the version of the product intended to be reused. Although a potentially reusable product exists, the lack of familiarity with the product may lead to the organization re-assessing the reusability of the product and alternative approach may be selected.

The reuse framework is similar to other process-based frameworks in that it provides prescriptive guidance for where, when, and how to apply the model. By identifying a standard, common process for how organizations should address, execute, and manage systems engineering reuse as well as a consistent approach for the application of COSYSMO 2.0, organizations will be able to derive greater, and more frequent, benefits from reuse. In addition to this guidance, the framework also highlights issues not addressed in the reuse model, such as the influence of other considerations not fully accounted for in the COSYSMO size drivers or COSYSMO 2.0 reuse categories and the potential impact of a reuse strategy that does not align with the products available for reuse.

Chapter 4: COSYSMO 2.0

4.1. Development of Model

COSYSMO is part of the Constructive Cost Model (COCOMO) family of cost models (Boehm 1981). Since COCOMO was first published, the model has led to the development of over ten related software and systems cost models at USC-CSSE alone, and more continue to be created. Both COSYSMO and COCOMO influenced the development of COSYSMO 2.0.

COSYSMO 2.0 is an extension of COSYSMO and the reuse model was developed to operate within the structural constraints of COSYSMO as well as also improve the estimation capabilities of the model. Despite the addition of the reuse extension, COSYSMO 2.0 does not alter the underlying parametric equation for COSYSMO, presented in Equation 2 (Valerdi 2005).

Equation 2 $PM = A * (Size)^E * (EM)$

Where:

PM = Person months

A = Calibration factor

$Size$ = Measure(s) of functional size of a system that has an additive effect on systems engineering effort

E = Scale factor(s) having an exponential or nonlinear effect on systems engineering effort

EM = Effort multipliers that influence systems engineering effort

Initial versions of COSYSMO 2.0 also drew heavily on the COCOMO (and then COCOMO II) reuse sub-model for guidance on incorporating the effect of reuse on effort estimates. COCOMO handles the reuse of software (also referred to as adapting existing software) by calculating the equivalent number of delivered source instructions, which is then used in place of the standard delivered source instructions (Boehm 1981). In 2000, COCOMO II was developed as an update to the original COCOMO model

(Boehm et al. 2000), which elaborated on the previous extension for a software reuse sub-model, presented in Equation 3.

Equation 3

$$AAF = 0.4(DM) + 0.3(CM) + 0.3(IM)$$

$$ESLOC = \frac{ASLOC[AA + AAF(1 + 0.02(SU)(UNFM))]}{100}, AAF \leq 0.5$$

$$ESLOC = \frac{ASLOC[AA + AAF + (SU)(UNFM)]}{100}, AAF > 0.5$$

Where:

ASLOC = Adapted Source Lines of Code

DM = Percent of Design Modification

CM = Percent of Code Modification

IM = Percent of Integration Required for Modified Software

SU = Percentage of reuse effort due to Software Understanding

AA = Percentage of reuse effort due to Assessment and Assimilation

UNFM = Programmer Unfamiliarity with Software

Variants of the COCOMO II factors, such as "unfamiliarity" and "understanding", were initially considered for incorporation into COSYSMO 2.0, but ultimately the model diverged from a COCOMO II-related structure due to the inability of software engineering-inspired reuse parameters to adequately map to the systems engineering domain. The modeling approach presented in the COCOMO II reuse sub-model of adjusting the size of the project through the application of various levels of reuse was determined relevant in systems engineering and a similar approach was utilized in COSYSMO 2.0.

4.2. Model Evolution

The initial methodology for a reuse extension to COSYSMO was to revise the estimate of the size of the systems engineering project by applying four types of reuse to the Number of Requirements size driver (Valerdi et al. 2006). This methodology divided

31

the three categories of Number of Requirements size driver, Easy, Nominal and Difficult, into four sub-categories, New, Modified, Reused, and Deleted requirements, defined in Table 8.

Reuse Category	Definition
1) New	Items that are completely new
2) Adopted	Items that are incorporated unmodified
3) Modified	Items that are reused but are tailored
4) Deleted	Items that are removed from a system

Table 8 Four Reuse Category Definitions

A model user would provide system specific quantities for each of the applicable sub-categories. For example, described in detail in Chapter 5.2.1, a model user may evaluate a system to have 185 New requirements (R_N), 60 Adopted requirements (R_A), 25 Modified requirements (R_O), and 0 Deleted requirements (R_D). The number of total requirements (R_T) would be calculated by multiplying the quantity for each sub-category by a pre-determined weighting factor (w_Y; where y = {New, Adopted, Modified, Deleted}) and then summing the products across the four categories, as shown in Equation 4.

Equation 4 $R_T = (w_Y \times R_Y)$

This calculation would be repeated across each of the Easy, Nominal, and Difficult rating categories for the Number of Requirements size driver. The outcome of these calculations would be a revised value for the Number of Requirements total, termed Total Equivalent New Requirements (R_{TE}), which would replace Number of Requirements in the model, as shown in Equation 5.

Equation 5 $R_{TE} = (w_N \times R_N) + (w_A \times R_A) + (w_O \times R_O) + (w_D \times R_D)$

The aim of this methodology was to reduce the number of requirements counted by COSYSMO by the amount of requirements to be reused. Although this methodology created a reasonable approach for accounting for reuse in the model (later strategies would build upon it), it was the first attempt at such a capability. Being the first attempt, however, a consensus could not be reached on the appropriate number and scope of reuse categories, and this approach did not receive full buy-in with the affiliate community as the acknowledged approach for reuse in COSYSMO (Wang 2007) (Valerdi, Wang et al. 2007). The significant contribution from this initial methodology was the categorization of non-new requirements as adopted, modified, or deleted.

Using these four reuse categories, a USC-CSSE affiliate created the COSYSMO Risk/Reuse model, known as COSYSMO-R, in 2007, a tool that provided organizations with a structure for applying the four reuse categories to COSYSMO, a screenshot of the model is presented in Figure 8 (Gaffney 2007). This was the first attempt at including a reuse estimation capability in a COSYSMO tool. With COSYSMO-R, users could assign weights for each of the reuse categories, which would revise the overall COSYSMO estimate.

New, Modified, Reused, and Deleted Requirements Proportions Data Entry.

	Easy				Nominal					Difficult				
Total [1]	New	Modified	Reused	Deleted	Total [1]	New	Modified	Reused	Deleted	Total [1]	New	Modified	Reused	Deleted
10	55.0%	10.0%	30.0%	5.0%	11	62.0%	5.0%	30.0%	3.0%	2	97.0%	0.0%	0.0%	3.0%
10	5.50	1.00	3.00	0.50	11	6.82	0.55	3.3	0.33	2	1.94	0.00	0.00	0.06
2	70.0%	5.0%	20.0%	5.0%	11	71.0%	2.0%	25.0%	2.0%	4	78.0%	1.0%	20.0%	1.0%
2	1.40	0.10	0.40	0.10	11	7.81	0.22	2.75	0.22	4	3.12	0.04	0.80	0.04
4	59.0%	5.0%	30.0%	6.0%	10	96.0%	4.00%	27.00%	3.00%	3	58.0%	2.00%	38.00%	5.00%
4	2.36	0.20	1.20	0.24	10	6.60	0.40	2.70	0.30	3	1.74	0.06	1.26	0.15
2	63.0%	4.00%	30.00%	3.00%	5	71.0%	3.00%	24.00%	2.00%	6	91.0%	3.00%	5.00%	1.00%
2	1.26	0.08	0.60	0.06	5	3.55	0.15	1.20	0.10	6	4.55	0.15	0.25	0.05

(1): The Total counts are the "Likely" size values on sheet COSYSMAIN.

Proportions [2]

	Pₙ	Pₘ	Pᵣ	Pᵈ	Check
# of System Requirements	74.12%	4.04%	18.46%	3.38%	100.00%
# of System Interfaces	73.99%	1.98%	22.65%	1.66%	100.00%
# of Algorithms	62.00%	3.29%	30.59%	4.13%	100.00%
# of Operational Scenarios	83.38%	3.05%	12.16%	1.41%	100.00%

Figure 8 COSYSMO-R Model Screenshot

After the publication of the initial reuse methodology and categories described above, a systems engineering organization applied it in an industrial setting (Wang, Valerdi et al. 2008). The outcome of this application indicated that while the four category reuse methodology was able to account for the effect of systems engineering reuse and did improve the estimation power of the model, four categories were inadequate at capturing all types of reuse. Specifically, an additional category was needed to capture instances where systems engineering products are reused without modification or testing (i.e. when a system integrator incorporates a product from a subcontractor). As a result, a second methodology was developed that included a fifth reuse category, presented in Table 9, to address instances of "managed" reuse (R_M). For the organization that applied this methodology, the five reuse categories further improved the estimation power of COSYSMO.

Reuse Category	Definition
1) New	Items that are completely new
2) Modified	Items that are incorporated unmodified
3) Adopted	Items that are reused but are tailored
4) Deleted	Items that are removed from a system
5) Managed	Items that are incorporated unmodified and untested

Table 9 Five Reuse Category Definitions

While validating this revised approach among systems engineering organizations, several issues were raised. First, the five categories fail to account for instances of "designed for reuse", when an upfront investment has been made to configure a systems engineering product so that it is reusable in anticipation of greater benefits throughout the life cycle. Second, the categories are only applied to a single COSYSMO size driver, Number of Requirements. Although the Number of Requirements driver contributes a significant amount to the estimation power of the model, reuse has an effect on the other drivers as well. Lastly, more reuse categories provide additional explanatory detail to the model, but also increase the difficulty of implementation. To address these issues, a third methodology was developed. This methodology expanded the number of reuse categories to six and established a structure of prime and sub-categories, presented in Table 10. For additional clarification and to reduce potential confusion, "items" in the category definitions were revised to "products" when the six categories were introduced.

Prime-Category	Sub-Category	Definition
I) New		Products that are completely new
	i) Designed for Reuse	Products that require an additional upfront investment to improve the potential reusability
II) Modified		Products that are inherited, but are tailored
	ii) Deleted	Products that are removed from a system
III) Adopted		Products that are incorporated unmodified (also known as "black box" reuse)
	iii) Managed	Products that are incorporated unmodified and with minimal testing

Table 10 Six Reuse Category Definitions

This third methodology addresses each of the issues previously raised. Although "designed for reuse" was known to occur, previous methodologies did not explicitly call it out as a separate reuse category, instead it was grouped into the New category. With a separate Designed for Reuse (R_F) category, the upfront investment for developing a reusable systems engineering product can be accounted for, which will typically exceed the cost of developing a new, non-reusable product, but may result in greater benefits (cost or effort savings) throughout the life cycle when reused. The prime/sub reuse category approach enables varying degrees of utilization of the reuse model. Some organizations expressed interest in applying only a few reuse categories, while others expressed interest in more categories and greater levels of detail. This approach is intended to capture three major, or prime, categories of reuse (New, Modified, and Adopted) as well as the three minor, or sub, categories of reuse (Designed for Reuse, Deleted, and Managed). The number of Total Equivalent New Requirements for a system with reuse can be calculated with Equation 6, a modified version of Equation 5, which includes the six reuse categories.

Equation 6

$$R_{TE} = \left(w_N \times R_N\right) + \left(w_F \times R_F\right) + \left(w_O \times R_O\right) + \left(w_D \times R_D\right) + \left(w_A \times R_A\right) + \left(w_M \times R_M\right)$$

This methodology can also be applied to not only the Number of Requirements size driver, but also the other three COSYSMO size drivers. Similar versions of the

Total Equivalent New Requirements equation are applicable to the Number of Interfaces, Number of Algorithms, and Number of Operational Scenarios.

An organization can use all or a subset of the six reuse categories when performing an estimate, depending on the particular circumstance. While more categories will likely produce a more comprehensive estimate, fewer categories will still enable accounting for some of the effect of reuse. These six reuse categories are the foundation for COSYSMO 2.0.

4.3. Model Form

Similar to COSYSMO and its reuse extension other models in the COCOMO family have been developed as extensions to COCOMO to better account for different types of software reuse (Boehm et al. 2005). Models such as COCOTS (Abts et al. 2001), COPLIMO (Boehm et al. 2004), and COSYSMO 2.0 extend the capabilities of their predecessors, while maintaining similar mathematical form. As a result, the number and scope of the size and cost drivers of COSYSMO remain the same in COSYSMO 2.0; the six reuse categories are added as an extension to the size drivers, but have no effect on the estimate if no reuse values are inputted. The operational concept of COSYSMO 2.0 is illustrated in Figure 9. Aside from the application of the reuse categories to the size drivers, the concept is the same for both models.

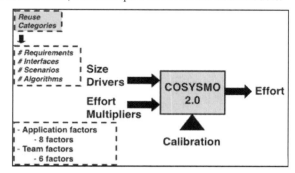

Figure 9 COSYSMO 2.0 Operational Concept

In the four COSYSMO size drivers, the user identifies the number of Easy, Nominal, and Difficult requirements, interfaces, operational scenarios, and system-specific algorithms. Using these as inputs and taking historical data into account, the model calculates the expected size of the system, in person-months, to complete the systems engineering activities for the particular system of interest (Valerdi 2005). For situations where reuse is significant, some amount of the systems engineering activities should not need to be completed and the expected amount of time to complete the remaining systems engineering activities should be less than what was originally estimated in COSYSMO. In COSYSMO 2.0, the estimate of system size is modified according to the weights of the reuse categories. For any requirement, interface, operational scenario, or system-specific algorithm being reused, the user still assigns it to an Easy, Nominal, or Difficult categories, and then determines which category of reuse is appropriate. The weight associated with each category of reuse is then multiplied across each of the three rating levels for each of the four size drivers. The result of this calculation is a revised estimate for project size, typically less than previously estimated by COSYSMO, and since reuse is accounted for, more representative of the realistic project effort.

The COSYSMO 2.0 operational equation is presented in

Equation 7. The difference between the COSYSMO and COSYSMO 2.0 operational equation is the addition of the second summation and the second weight parameter, which collectively account for the effect of reuse.

Equation 7

$$PM_{NS} = A \cdot \left[\sum_{k} \left(\sum_{r} w_r \left(w_{e,k} \Phi_{e,k} + w_{n,k} \Phi_{n,k} + w_{d,k} \Phi_{d,k} \right) \right) \right]^E \cdot \prod_{j=1}^{14} EM_j$$

Where:

PM_{NS} = effort in Person Months (Nominal Schedule)

A = calibration constant derived from historical project data

w_r = weight for reuse category

r = {New, Designed for Reuse, Modified, Deleted, Adopted, Managed}

$w_{x,k}$ = weight for size driver

x = {Easy, Nominal, Difficult}

k = {Requirements, Interfaces, Algorithms, Scenarios}

$\Phi_{x,k}$ = quantity of "k" size driver

E = represents (dis)economies of scale

EM = effort multiplier for the jth cost driver.

The second summation aggregates the weight of each reuse category multiplied across each size driver. The second weight parameter is the weight of each of the reuse categories. The New category has a weight of 1.0, since there is no reuse associated with it. The Modified, Deleted, Adopted, and Managed categories have weights less than 1.0 since there is some amount of reuse associated with them; a product with one of these levels of reuse should require less effort than a new, non-reused product. The Designed for Reuse category has a weight greater than 1.0 since some first-time investment needs to be made to make something reusable; however, subsequent uses of a systems

38

engineering product that was designed for reuse should be rated according to the expected reuse benefit and with a weight less than 1.0.

COSYSMO 2.0 accounts for reuse in the size drivers, rather than in the cost drivers, for three reasons. The first is that the four size drivers represent the majority of the explanatory power of the model, or more simply, the variability in a COSYSMO estimate is explained mostly by the size drivers. By accounting for reuse in the size drivers instead of the cost drivers, the effect of reuse will more directly influence the resulting estimate. Another reason for reuse being accounted for in the size drivers is that, to a limited degree, the availability of reusable tools is already accounted for in COSYSMO as part of the Tool Support cost driver. To avoid any confusion in double counting for the effect of reuse with multiple and/or overlapping cost drivers, the reuse model is limited to the size drivers. By keeping the number and scope of the size and cost drivers the same in both COSYSMO and COSYSMO 2.0, and having reuse be an extension to the model, backward compatibility between both models and their data is maintained. A third reason for implementing the reuse weights on the size drivers is that the size drivers are continuous variables, as opposed to the cost drivers, which are discrete variables. This provides the necessary level of fidelity and sensitivity to capture the impact of reuse on systems engineering size since there is a wider range of possible values that can exist to represent reuse. One of the original hypotheses of COSYSMO is that:

A combination of the four elements of functional size in [the model] contributes significantly to the accurate estimation of systems engineering (Valerdi 2005).

In other words, for a nominal project, the four size drivers adequately estimate the amount of effort in equivalent person-months that is required for a systems engineering project. As stated in Chapter 1.3, COSYSMO 2.0 follows a similar hypothesis:

Each COSYSMO size driver can be further decomposed into New, Designed for Reuse, Modified, Deleted, Adopted, and Managed categories of reuse each with corresponding rating scales and weights, and function as accurate predictors of equivalent size.

Following that the COSYSMO hypothesis is supported, the COSYSMO 2.0 hypothesis proposes that the application of the six reuse categories across all four size drivers adequately captures the effect of reuse on the amount of expected systems engineering effort.

Chapter 5: Methodology

5.1. Research Design and Data Collection

With a background on systems engineering reuse, the motivation for this research presented, and the reuse model proposed, the next steps in this research involve the identification of the appropriate research design and approach, and the correct application of these methods to the research question (Isaac and Michael 1997).

Designing a research methodology is not something that is solved, but rather it is a set of dilemmas that a researcher must decide how best to live with (McGrath 1981). The process of research design involves identifying multiple research methods and then evaluating the applicability and output of each method, sometimes iteratively. Every research method has both strengths and vulnerabilities, and it is very difficult for a single method to provide a researcher with a complete solution to a research problem. Because of this limitation, multiple types of research methods, also known as mixed-methods research, are often necessary to obtain convergence on a research result.

Mixed-methods research involves collecting, analyzing, and mixing one or more quantitative and qualitative approaches within a single research study (Creswell et al. 2006). A mixed-methods approach is relevant, and necessary, to conduct research on systems engineering reuse as it provides insight to different dimensions of reuse that otherwise could not be addressed by a single qualitative or quantitative research method. The overall research approach and specific research techniques are illustrated in Figure 10. The three hypothesis were evaluated, and eventually supported, by results from a variety of quantitative and qualitative research techniques including: literature reviews, a survey on the state of the practice, COSYSMO Workshop discussions, interviews with experts at systems engineering organizations, a Delphi survey, and an implementation of the model in an industrial setting.

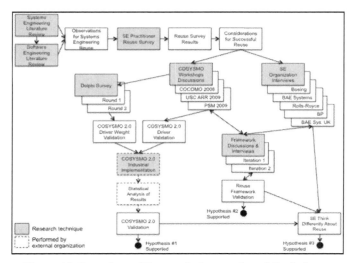

Figure 10 Overview of Research Approach and Techniques

Six research techniques were utilized to evaluate systems engineering reuse, an overview of these techniques was provided in Figure 10. First, a literature review was completed to assess the state of the art in both systems and software engineering reuse. The results of this review led to a series of observations that guided a practitioner survey. Second, a survey was developed and distributed to systems engineering practitioners to collect information about the state of the practice among systems engineering organizations. The concise, ten-question survey utilized a series of open and closed questions, and the results guided the development of the reuse model. Third, multiple COSYSMO Workshops were organized and attended by subject matter experts who discussed the various types of reuse and how their organizations manage reuse. Fourth, during these workshops, multiple rounds of a Wideband Delphi (Boehm 1981) survey were conducted. The Wideband Delphi exercise utilized discussions amongst a group of subject matter experts to produce quantitative weights for the reuse model parameters. Fifth, interviews and mini-case studies were held with multiple experts familiar with

42

reuse in their respective systems engineering organizations. The interviews were semi-structured, meaning that a set of questions was asked to each interviewee, but additional elaboration and discussion was encouraged. For industrial security reasons, the interviewees requested their responses not be attributable to respective organizations. These interviews collected information on the processes for how specific organizations handle reuse, which served as a validation of the reuse framework and were guided by the principles of case study design as described by Yin (2003). Lastly, a version of COSYSMO 2.0 was implemented at a large, diversified systems engineering organization across 44 projects and the results were obtained as validation.

Along with the overall mixed-methods approach, COSYSMO 2.0, as with COSYSMO, used both an interpretivist and positivist research approach (Klein and Myers 1999) to develop and propose the reuse model. The interpretivist approach was used to identify the reuse categories and preliminary definitions of the model. After COSYSMO 2.0 had been defined, the strategy shifted to a positivist approach. The positivist approach was used to make formal propositions, conduct hypothesis testing, and draw inferences about a phenomenon from a representative sample to a stated population.

The mixed-methods research approach used is similar to the research approaches utilized in the development of COSYSMO as well as other models in the COCOMO family, including COCOTS and COQUALMO (Baik et al. 2002). COSYSMO 2.0 followed the eight-step modeling methodology (Valerdi 2008), an updated version of the seven-step modeling methodology (Boehm et al. 2000), presented in Figure 11.

43

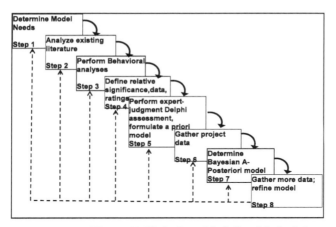

Figure 11 Eight-Step Modeling Methodology

This methodology involves: 1) identifying the need for accounting for reuse in COSYSMO, 2) analyzing existing literature for factors that affect systems engineering reuse, 3) conducting studies to determine how and under what conditions reuse occurs, 4) identifying the relative categories of reuse and their effect on systems engineering effort, 5) performing the Delphi assessment (Dalkey 1969) to validate the weights of the reuse categories, 6) gathering historical data on projects with reuse and determining statistical significance of the various parameters, 7) determining the Bayesian posterior set of model parameters, and 8) gathering more data to refine the model based on the experiences.

Initially this research was aimed at collecting historical data on systems engineering projects with reuse from aerospace and defense contractors. This data would be used to assess the ability of COSYSMO 2.0 to estimate projects with reuse and validate the improvement to the estimation power of COSYSMO. As with most systems engineering research, opportunities for data collection are limited (Valerdi and Davidz 2009) due to the size and length of most systems engineering projects, and the fact that many are developed for proprietary or military applications. Knowing this, systems engineering project data was still solicited from industry affiliates supporting this research. After several months of solicitation and discussion with subject matter experts, it was concluded that very limited data collection opportunities existed. Because systems engineering reuse was still an emerging concept, few organizations had collected historical information on reuse along with systems engineering effort data.

One organization was determined to have data on systems engineering reuse, although for industrial security reasons, the data could not be used or distributed externally. However, as explained later in Section 6.2, the organization did agree to implement a version of COSYSMO 2.0 using the model form and parameter weights derived from this research, and provide selected results. In light of this arrangement, parts of steps six through eight of the modeling methodology were performed by an external organization, under the guidance from this research. Due to the limited historical data collection opportunities, the results of this research were validated through a combination expert opinion, via a Delphi survey, and an implementation of the model across 44 projects at a diversified systems engineering organization.

Additionally, the existing COSYSMO calibration data set of 42 projects was re-examined for any potentially usable data on reuse. This analysis concluded a single reuse category was insufficient at capturing the effect of reuse and multiple reuse categories would be required; a summary of this analysis is described in Appendix E. Although access to actual project data was limited, the results of the Delphi survey and the industrial implementation of the model provided sufficient information to support the validation of the model.

With limited access to historical data, collecting expert opinion is the next most suitable means of validating research (Jorgensen and Boehm 2009). Since historical systems engineering reuse data has not been consistently collected, this research relied in part on empirical methods. The empirical method is to obtain an understanding through observations (Singleton and Straits 2009).

The empirical method is necessary in entering hitherto completely unexplored fields and becomes less purely empirical as the acquired mastery of the field increases (Bridgmann and Holton 2000).

Using this method, results from expert opinion surveys, indirect observations, expert interviews, and direct observations were aggregated from multiple sources, which improved the generalizability of the conclusions of this research. These different perspectives from multiple data sources are beneficial in the definition and testing of the hypotheses.

In summary, the question posed by this research was evaluated by data collected in the form of subject matter expert opinion from a practitioner survey, multiple COSYSMO Workshops, a Delphi survey, and several expert interviews. Data from the implementation of COSYSMO 2.0 at one organization was also collected. Table 11 illustrates how each of the research techniques and data collected support the testing of each hypothesis.

Research Technique	Hypothesis
Workshop discussions, Delphi survey, industrial data validation	#1: The requirements size driver can be further decomposed into New, Designed for Reuse, Modified, Adopted, Managed, and Deleted categories of reuse, each with corresponding definitions and weights, and function as good predictors of equivalent size from a systems engineering standpoint.
Literature review, practitioner survey, expert interviews and organizational validation	#2: By comparing the normative and descriptive models for managing systems engineering reuse, the systems engineering reuse framework provides a prescriptive model for enabling organizations to take advantage of reuse in systems engineering.
Workshop discussions, expert interviews, industrial data validation	#3: The COSYSMO 2.0 model and the associated systems engineering reuse framework help organizations think differently about systems engineering reuse, and its associated costs and benefits.

Table 11 Research Techniques to Hypothesis Mapping

5.2. Threats to Validity and Limitations

Threats to research validity can be grouped into either controllable or uncontrollable factors (Campbell and Stanley 1963). In this research, controllable validity issues are construct, convergent, internal, and external validity (McGrath 1981). Uncontrollable validity issues primarily deal with the use of subject matter expert opinion to generate results and validation of the model. This research identifies and addresses both controllable and uncontrollable validity issues.

Construct validity is when the parameters measure the intended construct and not something else. The construct in this research is the reuse of systems engineering products and it is measured through the number of new and reused requirements, interfaces, algorithms, and scenarios. The reuse of other systems engineering products, although not directly accounted for in reuse cost model, are considered to a degree in the reuse framework. Since the extent of systems engineering reuse can be subjective, the scope and definitions of the reuse categories attempt to ground this subjectivity in quantifiable terms when possible, but ultimately, values for the categories are surrogate measures of reuse. As qualitative measures, it is imperative that the intention and definition of the reuse categories in COSYSMO 2.0 be consistently interpreted and applied across organizations. Every attempt has been made to refine and clarify the scope of the reuse categories through COSYSMO Workshop discussions and interviews with subject matter experts. Since COSYSMO 2.0 is an extension of COSYSMO and uses a similar set of the model drivers and operational equation, the extension could have unintended consequences, potentially negatively impacting the estimation capability of COSYSMO. Given that the original COSYSMO defined eighteen systems engineering drivers and COSYSMO 2.0 sub-divides four of these drivers, there exists the potential for introducing new dependencies between the COSYSMO 2.0 reuse categories and the existing cost drivers. The scope and definitions of the reuse categories were crafted with the intention to minimize the overlap with existing drivers.

From a modeling perspective, a user can only account for the reuse of requirements, interfaces, algorithms, and scenarios. This may be constraining if the reuse of other systems engineering elements is expected. In practice, COSYSMO drivers should still be used to describe the aspects of systems without reuse and COSYSMO 2.0 drivers with the reuse categories should be used only to describe the characteristics of systems with reuse.

Convergent validity is when the research converges on common results. Using the mixed-methods approach, this research utilized multiple types of research techniques, listed in Table 11, to obtain result convergence. Due to the empirical nature of parts of this research, expert opinion was obtained from a several sources and methods in order to achieve a convergence of results. Although the reuse model was implemented at a single systems engineering organization, the quantitative results were compared and supported with qualitative results from expert opinion discussions.

Internal validity is the demonstration of a causal outcome between a treatment and an outcome. Internal validity is often difficult to demonstrate (Robson 2002), especially in systems engineering research, due to the complexity and breadth of confounding factors that influence the success of large-scale systems. This research addresses internal validity through iterations of the reuse category definitions at multiple workshop discussions, two rounds of a Delphi survey, an industrial implementation of the model, and interviews with experts on the framework; all of which illustrate the application of the reuse model and/or framework improve an organization's ability to estimate or manage reuse.

External validity is the generalizability of the research outside the context of the experiment. The results of the reuse model were validated across 44 projects at a diversified systems engineering organization, which participates in the aerospace and defense sector and is believed to be representative of similar organizations operating in related businesses. The applicability of these results to the commercial sector or smaller markets may not hold. The reuse framework was validated with interviews at multiple systems engineering organizations such as aerospace and defense contractors, as well as a natural resource exploration firm; therefore, the framework is expected to be applicable to a wide variety of systems engineering organizations.

While the above validity issues are mostly controllable, some other validity issues were mostly uncontrollable. Because of industrial security reasons, limited visibility existed in the implementation of the reuse model and this research relied on an external organization to follow the operational guidance and properly utilize the model. The quantitative analysis of these results assumed the implementation of the model was correct. This research also relied on expert opinion and subjective inputs. A researcher can be challenged to assess what characteristics constitute a true expert and whether an expert is biased. Although someone is identified as an expert, there is a limitation on the ability to screen for true experts. This issue was addressed by asking for background information on survey responders and utilizing discussion and interview results from personnel with previous experience in supporting COSYSMO research. Personnel who were identified as having the appropriate expertise, such as several years of experience practicing systems engineering and a familiarity with cost modeling, were invited to participate in the reuse survey and COSYSMO workshops.

Another mostly uncontrollable validity issue was the structural change to COSYSMO and the increased number of user inputs required for COSYSMO 2.0. The model increases the number of values a user must input into the size drivers from 12 to up to 60. Although most users will not be required to input 60 values into the model, the increase in the number from the COSYSMO model places an additional burden on the user, who must have a more detailed understanding of the system to utilize COSYSMO 2.0.

Along with these possible threats to validity, limitations also exist with the application of the model. Due to the limited historical reuse data, COSYSMO 2.0 was validated with data from one organization; an industry calibration from multiple data sources could not performed. Prior to utilizing the model for estimating, organizations will need to internally calibrate the model with historical data on completed projects. If data on projects with reuse do not exist, an organization should utilize experts familiar with the characteristics of their organization to make subjective assessments of past projects to generate calibration data similar to the methodology followed in this research. Since this research validated COSYSMO 2.0 but was unable to generate an industry calibration, the calibration range of the model may differ from the COSYSMO calibration range. An opportunity for future research is to collect reuse data on projects COSYSMO was calibrated to, which could improve the estimation capability of COSYSMO 2.0, although obtaining this reuse data for all of the projects in the existing COSYSMO data set is highly unlikely.

A final threat to validity is captured by the common business management phrase, "a fool with a tool is still a fool". When using the model, organizations need to be objective in their evaluation of reuse decisions and ensure the expected costs and benefits are compatible with their goals. With respect to the expect costs and benefits, as Boehm suggests to users of the COCOMO software reuse model, "be conservative" (1981).

Chapter 6: Results

6.1. Framework Validation

The reuse framework, introduced and described in Chapter 2.3, was created in response to a lack of an industry-accepted, standard process for how organizations manage systems engineering reuse, which was observed during the development of COSYSMO 2.0. Formulated through observations from systems and software engineering literature reviews, dialogue with subject matter experts, and discussions with multiple companies, the framework brings all aspects of the reuse issue together. Illustrated as a generic concept, the framework provides additional context to the reuse issue, documents prescriptive guidance for the operational usage of COSYSMO 2.0, and presents a repeatable process for how organizations can manage the reuse of systems engineering products. Although the framework assembles key information on reuse into a single illustration, it requires validation before it can be used and adopted by systems engineering organizations.

One methodology for validating a framework is to utilize industry examples, or case studies, to verify the concept and confirm the results. Researchers use case studies to illuminate decisions or sets of decisions within an individual, group, organizational, or social setting; generalize observations; and support theoretical propositions (Yin 2003). Case studies can be generated from documents, interviews, and observations. Although a single case study can be used to determine whether a proposition is correct, the results are much more convincing and have greater external validity if they are triangulated. Triangulation uses multiple case studies and several different sources of information to provide a convergence of evidence in support of a theory (Yin 2003). The triangulation method was utilized to validate the reuse framework through multiple sources: literature reviews, workshop discussions, interviews, and mini-case studies.

The framework was validated through discussions at two COSYSMO Workshops, with over a dozen companies represented, and interviews with multiple systems

51

engineering organizations, including BAE Systems U.S., BAE Systems U.K., Boeing, British Petroleum (BP), and Rolls-Royce. The two COSYSMO Workshops took place at the 2008 COCOMO Forum in Los Angeles, CA and the 2009 Practical Software and Systems Measurement (PSM) Conference in Orlando, FL. Between April 2009 and June 2009, three interviews and two mini-case studies were conducted. The workshop discussions and interviews guided the development of the framework; the results of the mini-case studies validated the framework. For industrial security reasons, the representatives of the mini-case studies requested the results to not be attributable to their organizations. For the purposes of this research, these two organizations are referred to as Organization A and Organization B.

At the COCOMO Forum in 2008, an initial iteration of the framework was presented to the COSYSMO Workshop attendees and discussed amongst the group. The questions and comments resulting from this discussion led to a second iteration of the framework, which received additional comments as it was presented to three companies during interviews. A final iteration of the framework was presented at the COSYSMO Workshop at the PSM Conference in 2009, with many participants from the first workshop in attendance. Discussions at this second workshop were very supportive of the framework and the attendees acknowledged it as a reasonable process for managing systems engineering reuse in their organizations and applying COSYSMO 2.0.

Following these workshops, mini-case studies with two systems engineering organizations were completed in order to assess how their organizations currently manage systems engineering reuse and if the framework adequately captured their current (or ideal) process.

In Organization A, systems engineering reuse occurs, but the organization does not have a highly mature process for implementing or managing it. The methodology for reuse in Organization A is focused around product centers. Organization A is comprised of multiple product centers, each responsible for multiple product families. Due to the

52

commonality of products within and across families, product centers are encouraged to leverage existing components or build new components to be applicable to multiple product needs. Software reuse is more common in Organization A than systems engineering reuse, but given the close relationship between software and systems engineering, advances in software reuse are forcing subsequent developments in systems engineering reuse. In the 1990's, Organization A primarily followed an opportunistic reuse strategy, when possible, reusing products that were previously developed and had a proven heritage. During this time, no formal process or plan existed for reuse and therefore, any reuse opportunities had to be identified by personnel with experience in the heritage product and insight into the new system. In the 2000's, Organization A observed customers were increasingly demanding custom solutions, but under constrained budgets and shorter development schedules. As these market forces placed competitive pressures on Organization A, reuse became a key means of satisfying the needs of the customer within budget and schedule constraints. Building reusability into products enabled Organization A to reuse certain products and focus instead on the integration of these components into a custom solution. The reuse framework was shown to the representative of Organization A, who stated that the framework is "applicable and valuable to organizations at multiple levels of reuse maturity". Although Organization A does not currently have a formal reuse process, the representative concluded the framework provides a basis for establishing a more formal process in Organization A and a consistent approach for applying COSYSMO 2.0.

In Organization B, a more mature reuse process exists compared to Organization A, but it is not consistently followed. Organization B does not frequently plan for reuse in the development of new systems and commonly follows an opportunistic reuse strategy; however, a process exists for archiving products into a reuse repository for future utilization. Organization B has internal goals for the amount of products reused and products contributed to the reuse repository. This reuse repository consists of the product, a record of when the product was reused, documentation on how the product was reused, and a rating. The rating is a measure determined by a committee of subject matter experts, that validates the applicability and usefulness of the product for reuse. The expert committee evaluates heuristics associated with the reuse of the product, the number of instances the product has been reused, and relevant metadata of the product. The rating process ensures products in the reuse repository are validated, thereby increasing the confidence of someone reusing the product. While Organization B considers the reuse repository to be a valuable tool, the representative noted that the additional resources required to review and rate products for reuse are frequently more than the expected benefit of reusing a product compared to building one from scratch. The reuse framework was presented to the representative of Organization B, who stated the framework is "a valid and reasonable illustration of the reuse process".

The results of the multiple workshop discussions, three interviews, and two mini-case studies supports the framework captures the reuse process and provides prescriptive guidance, and therefore validates the second hypothesis.

6.2. Model Validation

Due to the limited historical data on systems engineering reuse and few near-term data collection opportunities, subject matter expert opinion was used to derive the weights for the six reuse categories and the results from an implementation of a version of COSYSMO 2.0 was used to validate the model.

To derive the weights, two rounds of a Delphi survey were conducted at COSYSMO Workshops, one at the 2008 COCOMO Forum in Los Angeles, CA and one at the 2009 PSM Conference in Orlando, FL. The COSYSMO Workshops were selected as the venues for the survey for two reasons. First, one of the goals of each event was to guide cost model research. Second, both events had a reasonable number of attendees, 10-20 experts, and most with a moderate to advanced familiarity with COSYSMO as well as previous experience in supporting the development of the model. Participants in the Delphi survey included representatives from organizations such as The Aerospace Corporation, BAE Systems, Boeing, Lockheed Martin, and Northrop Grumman. The weight derivation exercise assessed which of the standard systems engineering activities, by life cycle phase, existed for each category of reuse.

The exercise, the format and results of which are presented in Appendix C and Appendix D, involved a review and discussion of each of the six reuse categories for each systems engineering activity by life cycle phase, to determine whether or not a particular activity would exist for that reuse category, in each life cycle phase. During the exercise, an expert would typically propose an assessment of a particular relationship, discussion among the experts would occur, and a consensus would be reached. If an activity relationship existed, an "X" was placed in a matrix where the activity, category, and life cycle phase intersect. If an activity relationship did not exist, the cell was left blank. With the Designed for Reuse category, "XX" represented instances where additional resources, greater than those for a product in rated in the New category, would be required to make a product reusable. At the COCOMO Forum, values for five of the six reuse categories (New, Modified, Deleted, Adopted, and Managed) were obtained. At the PSM Conference, values for the sixth reuse category (Designed for Reuse) were collected.

After completing the matrices in Appendix C and Appendix D, the weight of each reuse category was obtained by effectively adding the number of "X's" in each category. Previous research identified the standard effort associated with each activity and each life cycle phase (Valerdi and Wheaton 2005). By summing the activities with an "X" in each category, the percentage of effort for each category compared with the New category could be obtained following a "bottom up" approach (Wang, Valerdi and Fortune 2008). Since the New category was assumed to be the effort associated with completing systems engineering products without reuse, products with reuse categories of Modified, Deleted, Adopted, or Managed should have fewer activities (less effort)

55

comparatively, while products Designed for Reuse require additional time-consuming activities (more effort). The resulting calculated weights are presented in Figure 12 (Fortune 2009). These weights were then socialized among experts attending the COSYSMO Workshop and a consensus was obtained. Although the exercise was primarily focused on the effect of reuse on requirements, the results are believed to be applicable all COSYSMO size drivers and future research will further explore this applicability.

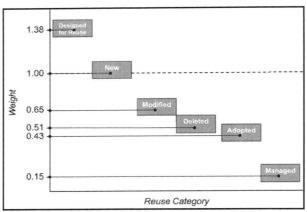

Figure 12 Reuse Category Weights

Overall, the outcomes of the reuse category weighting exercise aligned closely with the expected results. The Designed for Reuse category was confirmed to require a significant upfront investment, 38% beyond the cost of building a New product, in order to make a product reusable and realize greater reuse benefits in the future. The Managed category was determined to offer substantial reuse savings, although few products may actually fit within its definition. One limitation of the exercise results is that the weights in Figure 12 are point estimates, which represent the nominal reuse case, but not necessarily all reuse cases; some systems may have certain characteristics that influence the weights of the reuse categories beyond the nominal case. Future research will examine the possible ranges of these weights.

With the reuse model completed and weights for the parameters derived, a diversified systems engineering organization adopted the model and applied it across 44 systems engineering projects with reuse. The implementation utilized the definition of the reuse categories and weights of the parameters generated as products of this research, and the model was applied using five reuse categories across the Number of Requirements size driver. The five category version of COSYSMO 2.0, applied only to the Number of Requirements size driver, was implemented as a result of data availability. While a substantial amount of systems engineering reuse data did exist within the organization, data did not exist to populate the Designed for Reuse category. Furthermore, the reuse data was limited to the reuse of requirements, and therefore the model was only capable of being applied to the Number of Requirements size driver.

Due to industrial security reasons, only selected results of the COSYSMO 2.0 implementation could be disclosed as a significant amount of the model inputs and cost values were considered proprietary. As a result, some of the statistical analysis of the data was performed internally by the organization, but could not be released. The results that could be disclosed from the implementation of the model across the Number of Requirements size driver are presented in Figure 13. The graph on the top illustrates estimates of systems engineering effort with COSYSMO (with no reuse categories) and the graph on the bottom illustrates estimates of systems engineering effort with COSYSMO 2.0 (with five reuse categories).

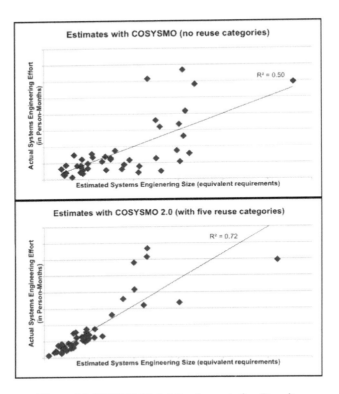

Figure 13 COSYSMO 2.0 Implementation Results

Using the available results, statistical analysis was performed to calculate the coefficient of determination, R^2, between the predictor, Estimated Systems Engineering Size, and the response (Weisberg 2005), Actual Systems Engineering Effort, and the predictive accuracy levels, PRED (Conte et al. 1986). With COSYSMO and no reuse categories applied, $R^2 = 0.50$ was produced; with COSYSMO 2.0 and five reuse categories applied, $R^2 = 0.72$ was produced. Therefore, the effort estimates from COSYSMO 2.0 with reuse had a higher correlation to the actual effort values than COSYSMO without reuse. Out of the 44 data points included in this analysis, COSYSMO 2.0 produced estimates closer to the actuals for 36 points, 82% of the set.

Table 12 compares the predictive accuracy of the two models to estimate within various ranges of the actual values. In most applications, COSYSMO estimates systems engineering effort within 30% of the actuals, 50% of the time, or PRED(.30) = 50%. For this application of COSYSMO using a set of data from a large, diversified systems engineering organization, the accuracy of the model was much worse than expected, possibly due to the projects involving a large degree of reuse. Using COSYSMO 2.0, the estimates were significantly improved. For a predictive accuracy range of 30% within the actuals, COSYSMO 2.0 estimates were 20% more accurate than COSYSMO estimates and even more accurate at higher accuracy ranges.

	Estimates with COSYSMO	Estimates with COSYSMO 2.0
PRED (.30)	14%	34%
PRED (.40)	20%	50%
PRED (.50)	20%	57%

Table 12 Statistical Comparison of Model Estimation Power

Campbell and Fiske state that multiple data points from a single organization are sufficient to support convergence on a conclusion and therefore validation (1959). With this implementation of COSYSMO 2.0 across 44 projects within a diversified organization, the improvement to the estimation power of COSYSMO through the COSYSMO 2.0 reuse extension is demonstrated. Applying COSYSMO 2.0 dramatically improved the power of the model and estimated systems engineering effort with better statistical accuracy than COSYSMO. These results validate COSYSMO 2.0 and support the first hypothesis.

6.2.1. Example COSYSMO 2.0 Estimate

With the model validated, an example estimate of a system with reuse using COSYSMO 2.0 can be provided. This example assumes a typical case in which a customer provides a system specification and requests an estimate of systems engineering effort from a contractor.

A customer has asked a contractor to estimate the amount of systems engineering effort that should be expected on an upcoming project. The customer provides the contractor with a system specification that contains 100 requirements (some reused), 10 interfaces (some reused), 7 system-specific algorithms, and 4 operational scenarios. Based on previous experiences with similar systems, the contractor has a high level of understanding of the requirements and there is a low level of technology risk.

Upon reviewing the provided data, the contractor determines there is sufficient data to utilize COSYSMO 2.0. After further review of the system specification, it is determined that the 100 requirements can be decomposed into 300 requirements at the systems engineering level. Through additional discussions with the customer, it is determined that of the 300 requirements at the systems engineering level; 100 are Easy, 150 are Nominal, and 50 are Difficult. Following discussions with other experts at the contractor, it is determined that of the 100 Easy requirements, 75 New and 25 are Modified; of the 150 Nominal requirements, 90 are New and 60 are Adopted; and of the 50 Difficult requirements, 20 are New and 30 are Adopted. A similar evaluation process is followed for the interfaces and system-specific algorithms. While reviewing the 10 interfaces, it is determined that 2 of the interfaces will be reused in the future and they should be designed for reuse. In order to make the interface reusable and generate a reuse benefit for future systems, an upfront investment is made during this system development. Out of the 10 interfaces, 8 are nominal and new, and 2 are difficult and will be designed for reuse. All of the 7 system-specific algorithms as well as the 4 operational scenarios are nominal and new.

These values are inputted into the COSYSMO 2.0 size drivers and reuse extension, and provide an initial person-month estimate for the systems engineering activities based solely on the systems engineering size parameters. Additional information about the system and the capabilities of the systems engineering team can further adjust this estimate. Since it was determined that the contractor has a high level of understanding of the requirements and a low level of technology risk exists for this system, the requirements understanding and technology risk cost drivers can be adjusted to high and low, respectively. These adjustments to the cost drivers further align the estimate with the expected capabilities of the systems engineering team.

Using the information from the system specification and supplementing it with customer and expert discussions, the COSYSMO 2.0 model inputs were populated and an estimate was generated. Based on the characteristics of this system and systems engineering team, COSYSMO 2.0 estimated 99.0 person-months to complete the systems engineering activities. An overview of the inputs and resulting estimate is provided in Figure 14. Comparatively, if the COSYSMO model (which does not account for the effect of reuse) was used instead of the COSYSMO 2.0 model, the estimate would have been 129.1 person-months, an overestimate of 30.4%.

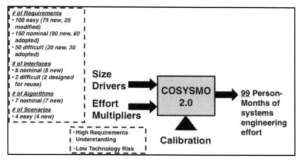

Figure 14 Example COSYSMO 2.0 Estimate

6.3. Implications for Systems Engineering Practice

Several implications for the systems engineering practice can be identified as a result of this research. These implications inform systems engineering practitioners about various aspects of reuse and offer guidance for managing reuse.

First, a formal process for evaluating, utilizing, and archiving reuse is necessary for successful reuse. A standard process brings consistency and reduces uncertainty across applications. A thorough and objective evaluation of the product to be reused and its intended application helps ensure that cost and benefit expectations are valid. A reuse repository captures information associated with a reuse success as well as a reuse failure, which could be extremely valuable in future reuse attempts. This implication is supported by the observations from the literature review and the interviews with experts about the reuse framework.

Second, a planned reuse strategy will almost always yield greater benefits and result in fewer issues than an opportunistic strategy. To maximize reuse opportunities and avoid potential architectural mismatches by introducing new product elements late in development, practitioners should assess and select products for reuse early in the life cycle. This implication is supported by the observations from the systems and software engineering literature review, and the COSYSMO Workshop discussions.

Third, reuse can dramatically influence the amount of effort associated with systems engineering activities and the costs and benefits of reuse should not be under or over-estimated. The results from the Delphi survey, the industrial implementation of COSYSMO 2.0, and even the example COSYSMO 2.0 estimate illustrate how significantly reuse can influence systems engineering effort. Practitioners should keep the expectations of the cost and benefit of reuse realistic; estimated benefits should be conservative and supported with historical data from related applications. While most systems engineering products can be reused, practitioners must realize that not all products can be reused efficiently; some upfront investment may be required before a product can be reused, which may exceed the cost of building a new product from scratch.

These implications and the following quotes from practitioners support the statement that this research "makes systems engineers think differently about reuse" and validate the third hypothesis.

> *Metrics-based estimates do account for the cost of systems engineering or systems engineering reuse very well. The parametric nature of COSYSMO 2.0, as COCOMO did for software reuse, improves fidelity and confidence in estimates of systems engineering reuse and provides systems engineers with valuable feedback on reuse decisions.*
> - Allan McQuarrie, Leveraged Technology & Products, BAE Systems

> *Systems Engineers are frequently tasked to leverage previously developed systems engineering products for the benefit of future projects. The inclusion of reuse in COSYSMO has an inherent value by adding clarity to project characterization and measurement refinement to the estimation of systems engineering cost.*
> - Marilee Wheaton, Systems Engineering Division, The Aerospace Corporation

> *The benefits of the COSYSMO reuse model are significant, including improving the accuracy of COSYSMO and demonstrating its validity across a wider set of systems engineering projects. The reuse model fundamentally enables the development of the Total Engineering Estimation Model and Process (TEEMaP) estimating model, an implementation of COSYSMO, to achieve high degrees of estimation accuracy and robustness. TEEMaP is now widely used in different Lines of Businesses.*
> - Gan Wang, Electronics and Integrated Solutions, BAE Systems

Chapter 7: Conclusion

7.1. Research Contributions to the Field of Systems Engineering

This research contributes to the fields of systems engineering in six key areas:

1) Identification of needs and gaps in the research of systems engineering reuse.

This research documented the deficiencies in the systems engineering literature and limitations in the collection of historical data on reuse. A review of the system and software engineering literature was performed, which influenced industry surveys and model development. Systems engineering practitioners were surveyed to identify existing definitions of reuse, products frequently reused, and expected criteria for successful reuse. The existence of systems engineering reuse was determined, but systems engineering organizations do not collect data on what, when, and where.

2) Definition and characterization of systems engineering reuse.

This research provides practitioners, researchers, and educators with definitions of systems engineering reuse as well as drivers that characterize the technical and non-technical considerations of reuse. Industry-validated definitions for systems engineering reuse products, categories, and strategies were established. The reuse of systems engineering products by systems engineering practitioners was documented. Technical and non-technical considerations that affect successful reuse, such as availability of existing products, horizontal organizational structures, experienced personnel, and formal documentation, were also documented. Agreement on the number, definition, and applicability of reuse categories was obtained through surveys, workshop discussions, and interviews with subject matter experts. The definitions and categories of reuse were approved by systems engineering organizations for future collection of systems engineering activity data.

3) Quantification of systems engineering reuse.

This research quantified the effect of various categories of reuse on systems engineering effort and validated COSYSMO 2.0, which will assist in the justification of expected costs and benefits associated with a reuse strategy. The limited historical data collected on reuse was identified as a limitation of the model, but validation by subject matter experts and the results of the implementation of the model at a diversified systems engineering organization are expected to be sufficient until a more comprehensive reuse data collection effort can be initiated. An analysis of the reuse data in the COSYSMO calibration data set concluded significant overestimation of projects exists and is attributable to reuse; furthermore, inconsistent definitions of reuse during the calibration data collection effort supported the need for multiple reuse categories. A methodology for quantifying and validating the reuse categories was developed, despite limitations in historical data.

4) Improvement of the COSYSMO tool to account for systems engineering reuse.

Practitioners benefit from this research through the development of a reuse extension for COSYSMO tool, as it improves their ability to estimate how reuse will affect the amount of expected systems engineering effort. This research established a modeling methodology of consistently accounting for reuse and at a level of granularity not previous available. COSYSMO 2.0 estimates the effect of reuse while balancing broad usability and direct applicability for systems engineering organizations.

5) Documentation of heuristics for systems engineering reuse.

This research benefits practitioners and educators by identifying potential systems engineering heuristics associated with systems engineering reuse; heuristics for systems engineering reuse were identified on the cost-benefit of reuse through the development of COSYSMO 2.0 and heuristics on risk, strategy, and product compatibility were identified through development of the reuse framework. The reuse framework captured the process for how systems engineering organizations should handle reuse, provided operational guidance for the COSYSMO 2.0 model, and documented aspects of reuse a cost model alone could not fully explain.

6) Systems engineers will think about their responsibilities differently.

65

Practitioners benefit from this research by being presented with a more complete perspective on the considerations associated with a reuse strategy and the effect to systems engineering effort. A framework was created for systems engineers to utilize a common process for acknowledging, planning, quantifying, assessing, and documenting reuse in their organizations. A process for formally recognizing reuse and pursuing a reuse strategy was established. A methodology to facilitate the consistent collection of data on reuse was identified, which improves future reuse opportunities. Best practices were provided for quantifying the effect of reuse with COSYSMO 2.0 and heuristics for assessing the associated benefits and risks. The need to archive products in a reuse repository to support future reuse opportunities was documented. The collection of the products generated by this research makes systems engineers think about reuse.

7.2. Areas for Future Research

This research presented a cost model for estimating systems engineering reuse and a framework for the reuse process. While these products are useful to systems engineers, their capabilities can be extended and be more useful through additional, long-term research. Opportunities for future research include:

1) Tailoring COSYSMO 2.0 to individual organizations.

Collecting additional historical data points will enable further refinement to the estimation power of the model as well as calibration of the model by organization and domain applications.

2) Exploring a range of weights for the reuse categories

Accounting for variation in the amount of reuse beyond point estimates will add further functionality to the model and provide model users with more granularity in developing estimates. Research on the range of reuse category weights would also include an examination of the effect of opportunistic versus planned reuse strategies on the weights.

3) Identifying reuse opportunities and mismatches.

Providing model users with criteria for identifying reuse opportunities or mismatches would increase the success of reuse strategies. Additional guidelines should be developed to highlight when issues may arise from a reused product being incongruent for an application domain or mismatched for a certain level of product specificity. Similar to the "no fly" zones of systems engineering identified by Bearden (2001).

4) Quantifying the effect of reuse on risk and schedule.

The model presented in this book focuses primarily on the cost and benefit effects of reuse; however, reuse also affects other programmatic considerations such as risk and schedule. Quantifying how reusing certain systems engineering products affect the development schedule and completion or operational risk of the systems would extend the applicability of the model and provide systems engineers with additional estimation capabilities.

5) Harmonizing reuse across hardware, software, and systems engineering.

COSYSMO 2.0 is limited to estimating the effect of systems engineering reuse. As research is being conducted to harmonize software and systems engineering estimation by combining elements of COCOMO and COSYSMO, harmonizing the effect of reuse across hardware, software, and systems engineering would be extremely valuable to systems engineering organizations.

Additional opportunities for research will be identified and created as the visibility of systems engineering reuse increases, and collaborations within and between industry and academia expand.

7.3. Lessons Learned for Systems Engineering Research

Conducting academic research in the field of systems engineering can be a daunting and challenging task for two reasons.

The first is the interaction between academic and industrial perspectives. Unlike many other research areas where academic and industrial research are decoupled, systems engineering research is primarily industry-focused and practitioner-driven. Understanding and leveraging this close relationship between systems engineering research and the systems engineering practice is a critical aspect of successfully completing research in this field. Some researchers may perceive the close relationship between systems engineering research and industry as a constraint, due to limitations associated with research goals, proprietary data, citizenship issues, or preference for near-term results. Because the major supporters of systems engineering research include aerospace, military, and large-scale commercial industries, research constraints do exist. However, this relationship should be embraced by researchers for the advantages it enables.

The second reason is the balance between various research methodologies necessary to make useful insights. Systems engineering projects frequently include both quantitative and qualitative decisions; therefore, a standard set of research techniques will not be applicable to every systems engineering research topic. A researcher will likely have to utilize a unique research methodology, incorporating multiple types of techniques, for each research project. Knowing this, systems engineering researchers should be familiar with both traditional engineering research techniques, as well as techniques from disciplines such as social science and psychology. Since systems engineering is a human-intensive activity it is necessary to integrate traditional engineering methods with social science methods of observation, such as surveys, interviews, and case studies. These techniques can assist both in the early stages of the research process, to solicit inputs and gather reach data about the phenomenon of interest, and the late stages of the process to validate results with expert opinion or real-world examples. Due to the diversity of challenges addressed by the systems engineering field, a researcher should be well versed in (and prepared to draw upon) research techniques from a variety of disciplines.

The systems engineering field is a relatively untapped research area with substantial demand for new research to be conducted, resources to support such research, and the ability to make an immediate contribution to the practice. Five specific recommendations for current and future systems engineering researchers should help move the maturity of the field a step forward:

Recommendation #1: A researcher should focus on identifying areas of opportunity that can be well-defined, fit within a narrow problem space (suitable for academic research).

Recommendation #2: Part of the topic identification process should include the consideration of possible research techniques that would be relevant to the research goals.

Recommendation #3: The topic of choice should be applicable enough to address the needs of practitioners, and achievable given the associated personnel, data, and time constraints.

Recommendation #4: There should be sufficient flexibility in the definition of hypotheses to allow for changes given data availability, fidelity and quality.

It is recommended that multiple hypotheses be defined so as to provide ample opportunities for data collection in different dimensions. From a risk mitigation standpoint, it is never good to put all of your research eggs in one basket in case the data you once counted on, or were promised, never materializes for reasons beyond your control.

Recommendation #5: Be prepared; data collection is the most difficult and time-consuming aspect of systems engineering research.

Testing and validating a hypothesis almost always requires data to be collected in one form or another. Since most systems engineering data resides within industrial organizations, researchers will likely need to obtain data that may be proprietary in nature or may be from multiple organizations, either of which may require data protection agreements. As a result, systems engineering researchers need to be cognizant of three key points.

First, it is the responsibility of the researcher to demonstrate their trustworthiness to industry organizations through strong data protection plans. Systems engineering data, because it is associated with so many sensitive aspects of a project, is very closely held. A researcher should not expect to be able to collect data until a relationship with the organization of interest has been developed and a data protection plan has been approved by the appropriate parties. The development of such a relationship and the release of data are not trivial tasks, in some cases taking months to years.

Sanitized data is frequently used in systems engineering research as a means of avoiding some of the sensitivity issues, but still providing adequate data to the researcher. If a researcher can work closely with the organization providing the sanitized data to ensure the proper collection procedures are followed and critical aspects of the data are not removed, this can be a very effective collection path. Once data is collected, a researcher must assess its quality before utilized; this is particularly true for systems engineering researchers, due to the multitude of interpretations of and mechanisms for accounting for systems engineering. Techniques such as examining metadata associated with the data of interest, comparing data from multiple sources within the same organization as well as different organizations, and utilizing clear, concise descriptions of data needs are all ways to improve data validity. Second, a researcher should be aware that because of the sensitivity of systems engineering data and the nature of the field being industry-focused, their employment affiliation will not always be beneficial. A researcher needs to be perceived as unbiased and objective by the community; being viewed as an employee rather than a researcher (professional or student) can limit opportunities for honest discussion and data collection. Ideally, a systems engineering researcher should be affiliated with (or emphasize their affiliation with) a university or other non-partisan research organization to minimize any conflict of interest or negative perception. The University Affiliated Research Center (UARC) for systems engineering is one example of a mechanism that enables researchers to be affiliated with such an organization, while still remaining connected to their home organization.

Third, and most importantly, prior to collecting data or even considering a research topic, a systems engineering researcher must persuade the industrial community that their research will produce a demonstrable, applicable, near-term benefit to the practice. Obtaining buy-in from practitioners on a research topic is critical as practitioners will be a source of research guidance, data collection opportunities, and validation of results. Opportunities for such interactions are available through participation in the Systems Engineering and Architecting Network for Research (SEANET), attendance at events sponsored by the International Council for Systems Engineering (INCOSE), and the presentation/publication of preliminary results at research forums such as the Conference on Systems Engineering Research (CSER).

Although the outlook for systems engineering research may sound grim, a wealth of opportunities does exist for systems engineering researchers, but it is not always obvious where. Researchers (or perspective researchers) should become active participants in systems engineering industry organizations, forums, and conferences. Not only do these venues provide a greater familiarity with the research areas of interest to the community, but also a means to solicit inputs, obtain guidance, and collect feedback from industry practitioners. A significant portion of systems engineering knowledge resides in these subject matter experts, which is not entirely captured in the academic literature. It is from interactions with industry organizations and practitioners that quality, actionable, and worthwhile research topics emerge.

The lessons learned presented in this section are intended to serve as guidance for both current and future systems engineering researchers on how to address and overcome some challenges associated with research in this field. As systems engineering continues to evolve, research priorities may change. However, the nature and diversity of the field will result in new research opportunities, which can be addressed by research synergies between academia and industry.

References

Abts, C., Boehm, B. W., and Bailey-Clark, E. (2001). "COCOTS : A Software COTS-Based System (CBS) Cost Model - Evolving Towards Maintenance Phase Modeling." Proceedings of the Twelfth Software Control and Metrics Conference, London, England.

Allen, T. Moses, J., Hastings, D., Lloyd, S., Little, J., McGowan, D., Magee, C., Moavenzadeh, F., Nightingale, D., Roos, D. and Whitney, D. (2001). Engineering Systems Division Terms and Definitions. Massachusetts Institute of Technology, Engineering Systems Division. Version 12.

ANSI/EIA (1999). ANSI/EIA-632-1988 Processes for Engineering a System.

Anthes, G. (1993). "Software Reuse Bring Paybacks." ComputerWorld. **Vol. 27** (No. 49).

Antelme, R., Moultrie, J. and Probert, D. (2000). "Engineering Reuse: A Framework for Improving Performance." IEEE International Conference on Management of Innovation and Technology, Singapore.

Baik, J., Boehm, B. W., and Steece, B. (2002). "Disaggregating and Calibrating the CASE Tool Variable in COCOMO II." IEEE Transactions on Software Engineering. **Vol. 28** (No. 11).

Basili, V., Romach, H., Bailey, J. and Joo, B. (1987). "Software Reuse: A Framework for Research." Tenth Minnowbrook Workshop on Software Performance Evaluation. Blue Mountain Lake, NY.

Basili, V. and Romach, H. (1991). "Support for Comprehensive Reuse." Software Engineering Journal. Vol. 6 (No. 5).

Bearden, D. (2001). "When is a Satellite Mission Too Fast and Too Cheap?" MAPLD International Conference, Laurel, MD.

Beckert, M. (2000). Organizational Characteristics for Successful Product Line Engineering. Masters Thesis. Massachusetts Institute of Technology. Cambridge, MA.

Blanchard, B. and Fabrycky, W. (1998). Systems Engineering and Analysis. Prentice Hall.

Boehm, B. W. (1981). Software Engineering Economics. Prentice Hall.

Boehm, B. W. and Scherlis, W. (1992). "Megaprogramming." Proceedings of the DARPA Software Technology Conference. (Available via USC Center for Systems and Software Engineering, Los Angeles, CA, TR-92-500).

Boehm, B. W. (1999). "Managing Software Productivity and Reuse." Computer. **Vol. 32** (No. 9).

Boehm, B. W., Abts, C., Brown, A. W., Chulani, S., Clark, B., Horowitz, E., Madachy, R., Reifer, D. J. and Steece, B. (2000). Software Cost Estimation With COCOMO II. Prentice Hall.

73

Boehm, B., Brown, W., Madachy, R., and Yang, Y. (2004). "A Software Product Line Life Cycle Cost Estimation Model." Proceedings of the 2004 International Symposium on Empirical Software Engineering. Redondo Beach, CA.

Boehm, B., Valerdi, R., Lane, J., and Brown, W. (2005). "COCOMO Suite Methodology and Evolution." CrossTalk. April 2005.

Bollinger, T. and Pfleeger, S. (1992). "Economics of Reuse: Issues and Alternatives." Information and Software Technology. **Vol. 32** (No. 10).

Budros, A. (1999). "A Conceptual Framework for Analyzing Why Organizations Divide." Organization Science. **Vol. 10** (No. 1).

Campell, D. and Fiske, D. (1959). "Convergent and Discriminant Validation by the Multitrait-Mulitmethod Matrix." Psychological Bulletin. **Vol. 56**.

Campbell, D. and Stanley, J. (1963). Experimental and Quasi-Experimental Designs for Research. Wadsworth Publishing.

Carlile, P. (2004). "Transferring, Translating, and Transforming: An Integrative Framework for Managing Knowledge across Boundaries." Organization Science. **Vol. 15** (No. 5).

Conte, S. D., Dunsmore, H. E., Shen, V. Y., (1986) Software Engineering Metrics and Models. Benjamin/Cummings Publishing Company.

Cooper, L., Majchrzak, A. and Faraj, S. (2005). "Learning from Project Experiences Using a Legacy-Based Approach." 38th Annual Hawaii International Conference on System Science. Big Island, HI.

Creswell, J., Shope, R., Plano Clark, V., and Green, D. (2006). "How Interpretive Qualitative Research Extends Mixed Methods Research." Research in the Schools. **Vol. 13** (No. 1).

Cybulski, J., Neal, R., Kram, A. and Allen, J. (1998). "Reuse of Early Life Cycle Artifacts: Workproducts, Methods, and Tools." Annals of Software Engineering. **Vol. 5** (No. 1).

Dalkey, N. (1969). The Delphi Method: An Experimental Study of Group Opinion. RAND Corporation.

Department of Defense. (2004). "Testing in Joint Environment Roadmap." Operational Test and Evaluation. Final Report.

Dusink, L and van Katwijk, J. (1995). "Reuse Dimensions." Symposium on Software Reliability. Seattle, WA.

Finkelstein, A. (1988). "Re-use of Formatted Requirements Specifications." Software Engineering Journal. **Vol. 3** (No. 5).

Fischer, J. (1971). "A Framework for the Analysis and Comparison of Clinical Theories of Induced Change." The Social Service Review. **Vol. 45** (No. 4).

Fortune, J. and Valerdi, R. (2008). "Considerations for Successful Reuse in Systems Engineering." AIAA Space 2008, San Diego, CA.

Fortune, J., Valerdi, R. and Wang, G. (2008). "Systems Engineering Reuse: A Report on the State of the Practice." 23rd International Forum on COCOMO and Systems/Software Cost Modeling, Los Angeles, CA.

Fortune, J. (2009). "COSYSMO Workshop Out-brief." Practical Systems and Software Measurement Conference, Orlando, FL.

Frakes, W. and Fox, C. (1995). "Sixteen Questions About Software Reuse." Communications of the ACM. **Vol. 38** (No. 6).

Gaffney, J. (2007). "COSYSMO-Risk/Reuse Model." Lockheed Martin.

Garlan, D., Allen, R. and Ockerbloom, J. (1995). "Architectural Mismatch: Why reuse is so hard." IEEE Software. **Vol. 12** (No. 6).

Garud, R. and Kumaraswamy, A. (1995). "Technological and Organizational Designs for Realizing Economies of Substitution." Strategic Management Journal. **Vol. 16**.

Glass, R. (1999). "Reuse: What's Wrong With This Picture?" IEEE Software. **Vol. 15** (No. 2).

IEEE. (1999). IEEE 1517-1999 – Software Life Cycle-Reuse Processes.

IEEE. (2005). IEEE 1220-2005 Systems Engineering – Application and Management of the Systems Engineering Process.

INCOSE. (2007). Systems Engineering Handbook. INCOSE. Version 3.1.

Isaac, S. and Michael, W. B. (1997). Handbook in Research and Evaluation. EdITS.

ISPA. (2007). Parametric Estimation Handbook. ISPA. Fourth Edition.

ISO/IEC (2002). ISO/IEC 15288:2002(E) Systems Engineering - System Life Cycle Processes.

Isoda, S. (1996). "Software Reuse in Japan." Information and Software Technology. **Vol. 38** (No. 3).

Jorgensen, M. and Boehm, B. (2009). "Software Development Effort Estimation: Formal Models or Expert Judgement?" IEEE Software. **Vol. 26** (No. 2).

de Judicibus, D. (1996). "Reuse: A Cultural Change." International Workshop on Systematic Reuse, Liverpool, England.

75

Khadilkar, D. and Stauffer, L. (1996). "An Experimental Evaluation of Design Information Reuse During Conceptual Design." Journal of Engineering Design. **Vol. 7** (No. 4).

Kitchenham, B., Pfleeger, S., and Fenton, N. (1995). "Towards a Framework for Software Measurement Validation." IEEE Transactions on Software Engineering. **Vol. 21** (No. 12).

Klein, H. K. and Myers, M. D. (1999). "A Set of Principles for Conducting and Evaluating Interpretive Field Studies in Information Systems." MIS Quarterly. **Vol. 23** (No. 1).

Konito, J., Caldiera, G. and Basili, V. (1996). "Defining Factors, Goals, and Criteria for Reusable Component Evaluation." Conference of the Centre for Advanced Studies on Collaborative Research. Toronto, Ontario, Canada.

Lam, W., McDermid, A. and Vickers, A. (1997). "Ten Steps Towards Systematic Requirements Reuse." Requirements Engineering. **Vol. 2** (No. 2).

Lam, W. and Loomes, M. (1998). "Re-engineering for Reuse: A Paradigm for Evolving Complex Reuse Artifacts." 22nd International Computer Software and Application Conference, Vienna, Austria.

Larreche, J. and Montgomery, D. (1977). "A Framework for the Comparison of Marketing Models: A Delphi Study." Journal of Marketing Research. **Vol. 14** (No. 4).

Lim, W. (1996). "Reuse Economics: A Comparison of Seventeen Models and Directions for Future Research." Fourth International Conference on Software Reuse. Orlando, FL.

Lim, W. (1998). Managing Software Reuse. Prentice Hall.

Lougee, H. (2004). "Reuse and DO-178B Certified Software: Beginning with Reuse Basics." CrossTalk. **Vol. 17** (No. 12).

Maier, M. and Rechtin, E. (2002). The Art of Systems Architecting. CRC Press.

Maier, M. (2006). "System and Software Architecture Reconciliation." Systems Engineering. **Vol. 9** (No. 2).

Malhotra, A. and Majchrzak, A. (2004). "Enabling Knowledge Creation in Far-Flung Teams: Best Practices for IT Support and Knowledge Sharing." Journal of Knowledge Management. **Vol. 8** (No. 4).

McGrath, J. (1981). "Dilemmatics: The Study of Research Choices and Dilemmas." American Behavioral Scientist. **Vol. 25** (No. 2).

Mili, H., Mili, A., Yacoub, S. and Addy, E. (2002). Reuse-Based Software Engineering. John Wiley & Sons.

Moore, M. (2001). "Software Reuse: Silver Bullet?" IEEE Software. **Vol. 18** (No. 5).

NASA. NASA Procedural Requirement 7123.1A - NASA Systems Engineering Processes and Requirements. 2007.

NASA. NASA Systems Engineering Handbook. Rev. 1. 2007.

Poulin, J. (1997). Measuring Software Reuse. Addison-Wesley.

Poulin, J. and Caruso, J. (1993). "Determining the Value of a Corporate Reuse Program." 1st International Software Metrics Symposium, Baltimore, MD.

Prieto-Diaz, R. (1993). "Status Report: Software Reusability." IEEE Software. **Vol. 10** (No. 3).

Prieto-Diaz, R. (1996). "Reuse as a New Paradigm for Software Development." International Workshop on Systematic Reuse, Liverpool, England.

Reifer, D. (1997). Practical Software Reuse. John Wiley & Sons.

Rieff, J., Gaffney, J. and Roedler, G. (2007). "2007: The Breakout Year for COSYSMO." PSM Users Group Conference, Golden, CO.

Robertson, S. (1996). "Reuse Lifecycle: Essentials and Implementations." International Workshop on Systematic Reuse, Liverpool, England.

Robson, C. (2002). Real World Research: A Resource for Social Scientists and Practitioner-Researchers. Wiley-Blackwell.

Sage, A. P. (1992). Systems Engineering, John Wiley & Sons, Inc.

Selby, R. (2005). "Software Reuse in Large-Scale Systems." AIAA Space 2005, Long Beach, CA.

Selby, R. (Ed.) (2007). Software Engineering. Wiley-Interscience.

Singleton, R. and Straits, B. (2009). Approaches to Social Research. Oxford University Press.

Stephens, R. (2004). "Measuring Enterprise Reuse in a Large Scale Corporate Environment." 37th Southeastern Symposium on System Theory. Fort Lauderdale, FL.

Szulanski, G. and Winter, S. (2002). "Getting It Right the Second Time." Harvard Business Review. **Vol. 80** (No. 1).

Tracz, W. (1988). "Software Reuse Myths." Software Engineering Notes. **Vol. 13** (No. 1).

Tracz, W. (1995). "Confessions of a Used Program Salesman: Lessons Learned." Symposium on Software Reusability. Seattle, WA.

Valerdi, R. (2005). The Constructive Systems Engineering Cost Model. Ph.D. Dissertation. University of Southern California. Los Angeles, CA.

Valerdi, R. and Wheaton, M. (2005). "ANSI/EIA 632 As a Standard WBS for COSYSMO." AIAA 1st Infotech at Aerospace Conference, Arlington, VA.

Valerdi, R., Gaffney, J., Roedler, G. and Rieff, J. (2006). "Extensions to COSYSMO to Represent Reuse." 21st International Forum on COCOMO and Software Cost Modeling, Los Angeles, CA.

Valerdi, R., Rieff, J., Roedler, G. and Wheaton, M. (2007). "Lessons Learned from Industrial Validation of COSYSMO." 17th INCOSE Symposium, June 2007, San Diego, CA.

Valerdi, R., Ross, A., and Rhodes, D. (2007). "A Framework for Evolving System of Systems Engineering." CrossTalk. October 2007.

Valerdi, R., Wang, G., Roedler, G., Rieff, J. and Fortune, J. (2007). "COSYSMO Reuse Extension." 22nd International Forum on COCOMO and Systems/Software Cost Modeling, Los Angeles, CA.

Valerdi, R. (2008). The Constructive Systems Engineering Cost Model (COSYSMO): Quantifying the Costs of Systems Engineering Effort in Complex Systems. VDM Verlag.

Valerdi, R., Roedler, G. and Fortune, J. (2008). "Towards COSYSMO 2.0: Future Directions and Priorities." USC-CSSE Annual Research Review, Los Angeles, CA.

Valerdi, R. and Davidz, H. (2009). "Empirical Research in Systems Engineering: Challenges and Opportunities of a New Frontier." Systems Engineering. Vol. 12 (No. 2).

Wang, G. (2007). "COSYSMO Extension: Reuse." 2007 COCOMO Forum. COSYSMO Working Group, Los Angeles, CA.

Wang, G., Valerdi, R., Ankrum, A., Millar, C. and Roedler, G. (2008). "COSYSMO Reuse Extension." 18th INCOSE Symposium, Utrecht, the Netherlands.

Wang, G., Valerdi, R. and Fortune, J. (2009). "Reuse in Systems Engineering." (Under Review), IEEE Systems.

Weisberg, S. (2005). Applied Linear Regression. John Wiley & Sons.

Whittle, B., Lam, W. and Kelly, T. (1996). "A Pragmatic Approach to Reuse Introduction in an Industrial Setting." Workshop on Systematic Reuse, Liverpool, UK.

Wiles, E. (1999). Economic Models of Software Reuse: A Survey, Comparison, and Partial Validation. Ph.D. Dissertation. University of Wales. Aberystwyth, Ceredigion, UK.

Wymore, A. and Bahill, A. (2000). "Can We Safely Reuse Systems, Upgrade Systems, or Use COTS Components?" Systems Engineering. Vol. 3 (No. 2).

Yin, R. (2003). Case Study Research Designs and Methods. Sage Publications. Third Edition.

Young, M. (1992). "A Framework for Successful Adoption and Performance of Japanese Manufacturing Practices in the United States." The Academy of Management Review. **Vol. 17** (No. 4).

Appendices

Appendix A: ANSI/EIA 632 Standard Systems Engineering Activities List

Acquistion and Supply	Supply Process	(1) Product Supply
	Acquisition Process	(2) Product Acquisition
		(3) Supplier Performance
Technical Management	Planning Process	(4) Process Implementation Strategy
		(5) Technical Effort Definition
		(6) Schedule and Organization
		(7) Technical Plans,
		(8) Work Directives
	Assessment Process	(9) Progress Against Plans and Schedules
		(10) Progress Against Requirements
		(11) Technical Reviews
	Control Process	(12) Outcomes Management
		(13) Information Dissemination
System Design	Requirements Definition Process	(14) Acquirer Requirements
		(15) Other Stakeholder Requirements
		(16) System Technical Requirements
	Solution Definition Process	(17) Logical Solution Representations
		(18) Physical Solution Representations
		(19) Specified Requirements
Product Realization	Implementation Process	(20) Implementation
	Transition to Use Process	(21) Transition to use
Technical Evaluation	Systems Analysis Process	(22) Effectiveness Analysis
		(23) Tradeoff Analysis
		(24) Risk Analysis
	Requirements Validation Process	(25) Requirement Statements Validation
		(26) Acquirer Requirements
		(27) Other Stakeholder Requirements,
		(28) System Technical Requirements
		(29) Logical Solution Representations
	System Verification Process	(30) Design Solution Verification
		(31) End Product Verification
		(32) Enabling Product Readiness
	End Products Validation Process	(33) End products validation

Table A-1 ANSI/EIA 632 Standard Systems Engineering Activities

COSYSMO 2.0 Reuse Survey

Description

The purpose of this survey is to collect information on the reuse of systems engineering artifacts from a previously developed system and determine how the reuse of such artifacts influences subsequent systems engineering activities and effort. The responses to this survey will be used to develop the reuse extension of the second iteration of the Constructive Systems Engineering Cost Model (COSYSMO 2.0), currently under development by the Center for Systems and Software Engineering at USC. This survey is being circulated to the Center for Systems and Software Engineering's affiliate organizations, members of the MIT Lean Advancement Initiative consortium, and other interested parties.

This research intends to identify, assess, and quantify reuse in the context of systems engineering. The study focus is limited to the reuse of systems engineering artifacts such as requirements, documentation, test procedures, knowledge, processes, etc. that influence systems engineering activities (as described by EIA/ANSI 632). Other reuse elements involving hardware or software are considered outside the scope of this study.

If any of your responses require you to make assumptions, please include those assumptions. Please submit forms via e-mail to:

Jared Fortune
University of Southern California
Center for Systems and Software Engineering
Email: fortune@usc.edu
Phone: 310-292-4087

81

1. General Information

1.1. Date

1.2. Company

1.3. Organization/Business Unit

1.4. Name of Responder

1.5. Email

1.6. Phone

1.7. Primary System Experience (check all that apply)

☐ Agriculture	☐ Aircraft/Avionics (Commercial jets, helicopters, avionics devices)	☐ Automotive / Motor Vehicles (cars, trucks, buses, etc.)
☐ Data Systems/Information Technology (health care, legal, business records and databases, etc.)	☐ Energy (coal, gas, oil, electric production and distribution, etc.)	☐ Environmental/Waste Mgt (restoration, preservation, conservation, waste mgt, etc.)
☐ Financial	☐ Geographic Information	☐ Infrastructure (Facilities, urban planning, asset mgt, etc.)
☐ Manufacturing	☐ Marine (Boats, ships, subs, etc)	☐ Medical Technology (Medical systems, devices, treatments)
☐ Military/Defense (Tanks, Missiles, etc.)	☐ Natural Resource Management (Water, etc.)	☐ Pharmaceutical/Chemical
☐ Scientific/Research	☐ Space Systems	☐ Telecommunications
☐ Transportation Systems (Railway, Air traffic, Highway, Waterway, etc.)	☐ Other	

2. Questions

2.1.   In the context of systems engineering, how does your organization define reuse? If a specific definition exists, please identify the document.

_____   _____

_____

_____

2.2.   What are the systems engineering artifacts that your organization reuses and how frequently are the artifacts reused?

		Never		Occasionally	Always	
a.	Requirements	1 ☐	2 ☐	3 ☐	4 ☐	5 ☐
b.	Documentation/Templates	1 ☐	2 ☐	3 ☐	4 ☐	5 ☐
c.	Test Data/Procedures	1 ☐	2 ☐	3 ☐	4 ☐	5 ☐
d.	Architecture/Design Models	1 ☐	2 ☐	3 ☐	4 ☐	5 ☐
e.	COTS Products (identify below)	1 ☐	2 ☐	3 ☐	4 ☐	5 ☐
	_____					
f.	Other (describe below)	1 ☐	2 ☐	3 ☐	4 ☐	5 ☐
	_____					
g.	Other (describe below)	1 ☐	2 ☐	3 ☐	4 ☐	5 ☐
	_____					

2.3.   In you opinion, which of the artifacts listed above is the most effective at providing a net benefit when reused?

_____

2.4.   In the systems lifecycle phases below, to what extent did the reuse of systems engineering artifacts occur? Select appropriate level.

		Unaware		Ad Hoc		Planned
a.	Conceptualize	1 ☐	2 ☐	3 ☐	4 ☐	5 ☐
b.	Develop	1 ☐	2 ☐	3 ☐	4 ☐	5 ☐
c.	Test and Evaluation	1 ☐	2 ☐	3 ☐	4 ☐	5 ☐
d.	Transition to Operation	1 ☐	2 ☐	3 ☐	4 ☐	5 ☐

2.5.   What have been the reuse successes? What have been the major reasons for reuse successes?

83

_____ _____
_____
_____

2.6. What have been the reuse failures? What have been the major reasons for reuse <u>failures</u>?

_____ _____
_____
_____

2.7. What are the most frequently promoted benefits as justification for systems engineering reuse? Rank from 1 to 6; 1 is most common, 6 is least common.

   a.   Cost

   b.   Schedule

   c.   Performance/productivity

   d.   Risk

   e.   Quality

   f.   Other (describe below)
        _____

2.8. How frequently is systems engineering reuse mentioned in a RFP for a new system? Choose one.

   a.   Never   ☐

   b.   Seldom   ☐

   c.   Occasionally   ☐

   d.   Always   ☐

2.9. In your opinion, how do the expected savings from reusing systems engineering artifacts scale (e.g. linear, non-linear) with system complexity or size?

_____   _____

_____

_____

2.10. Below are the five proposed categories of systems engineering reuse and their definitions:

New   Artifacts that are completely new.

Modified   Artifacts that are inherited from previous systems, but are tailored for the new system.

Adopted   Artifacts that are incorporated unmodified, "black box" reuse.

Deleted   Artifacts that are removed from a system.

Managed   Artifacts that are incorporated unmodified and untested.

For each pair (a through j) of reuse categories, consider the reuse of a systems engineering artifact, such as requirements. Evaluate the expected effort from utilizing a systems engineering artifact classified in the first reuse category compared to the second reuse category (e.g. "the expected effort from using a New requirement is Less Than / About the Same / More Than the expected effort from using an Modified requirement").

a.	New	☐ Less Than	☐ About the Same	☐ Greater than	Modified
b.	New	☐ Less Than	☐ About the Same	☐ Greater than	Adopted
c.	New	☐ Less Than	☐ About the Same	☐ Greater than	Deleted
d.	New	☐ Less Than	☐ About the Same	☐ Greater than	Managed
e.	Modified	☐ Less Than	☐ About the Same	☐ Greater than	Adopted
f.	Modified	☐ Less Than	☐ About the Same	☐ Greater than	Deleted
g.	Modified	☐ Less Than	☐ About the Same	☐ Greater than	Managed
h.	Adopted	☐ Less Than	☐ About the Same	☐ Greater than	Deleted
i.	Adopted	☐ Less Than	☐ About the Same	☐ Greater than	Managed
j.	Deleted	☐ Less Than	☐ About the Same	☐ Greater than	Managed

ISO/IEC 15288-Based Life Cycle Phases

Reuse Categories

IEA 632-Reuse Activity Cross Walk	SE Activities For NEW	SE Activities For MODIFIED	SE Activities For DELETED	SE Activities For ADOPTED	SE Activities For MANAGED

Acquisition and Supply
1. Product Supply
2. Product Acquisition
3. Supplier Performance

4. Process Implementation Strategy
5. Technical Effort Definition
6. Schedule and Organization
7. Technical Plans

Technical Management
8. Work Directives
9. Progress Against Plans and Schedules
10. Progress Against Requirements
11. Technical Reviews
12. Outcomes Management
13. Information Dissemination

System Design
14. Acquirer Requirements
15. Other Stakeholder Requirements
16. System Technical Requirements
17. Logical Solution Representations
18. Physical Solution Representations
19. Specified Requirements

Product Realization
20. Implementation
21. Transition to Use

Technical Evaluation
22. Effectiveness Analysis
23. Tradeoff Analysis
24. Risk Analysis
25. Requirements Statements Validation
26. Acquirer Requirements Validation
27. Other Stakeholder Requirements Validation
28. System Technical Requirements Validation
29. Logical Solution Representations Validation
30. Design Solution Verification
31. End Product Verification
32. Enabling Product Readiness
33. End Products Validation

Table C-1 Delphi Survey Results – Round 1

Appendix D: Reuse Category Delphi – Round 2 Results

ISO/IEC 15288-Based Life Cycle Phases		Conceptualize	Develop	Operate Test & Eval	Transition to Operation
		Reuse Categories			
	EIA 632-Reuse Activity Cross Walk	SE Activities For DESIGN FOR REUSE			
Acquisition and Supply	1. Product Supply	X	X	X	X
	2. Product Acquisition	X	X	X	X
	3. Supplier Performance	X	X	X	X
Technical Management	4. Process Implementation Strategy	X	X	X	X
	5. Technical Effort Definition	X	X	X	X
	6. Schedule and Organization	X	X	X	X
	7. Technical Plans	XX	XX	XX	X
	8. Work Directives	XX	XX	XX	X
	9. Progress Against Plans and Schedules	X	X	X	X
	10. Progress Against Requirements	XX	XX	X	X
	11. Technical Reviews	X	X	X	X
	12. Outcomes Management	XX	XX	X	X
	13. Information Dissemination	XX	XX	X	X
System Design	14. Acquirer Requirements	X	X	X	X
	15. Other Stakeholder Requirements	XX	XX	X	X
	16. System Technical Requirements	XX	XX	X	X
	17. Logical Solution Representations	XX	XX	X	X
	18. Physical Solution Representations	XX	XX	X	X
	19. Specified Requirements	XX	XX	X	X
Product Realization	20. Implementation	XX	XX	X	X
	21. Transition to Use	X	X	X	X
Technical Evaluation	22. Effectiveness Analysis	XX	XX	X	X
	23. Tradeoff Analysis	XX	XX	X	X
	24. Risk Analysis	XX	XX	X	X
	25. Requirements Statements Validation	XX	XX	X	X
	26. Acquirer Requirements Validation	X	X	X	X
	27. Other Stakeholder Requirements Validation	XX	XX	X	X
	28. System Technical Requirements Validation	XX	XX	X	X
	29. Logical Solution Representations Validation	XX	XX	X	X
	30. Design Solution Verification	XX	XX	XX	X
	31. End Product Verification	X	X	X	X
	32. Enabling Product Readiness	X	X	X	X
	33. End Products Validation	X	X	X	X

Table D-1

Delphi Survey Results – Round 2

Appendix E: COSYSMO Calibration Data Set Reuse Analysis

The existing COSYSMO calibration data set is the collection of systems engineering projects that were used to calibrate and validate the COSYSMO tool when it was published in 2005. Data on these projects were obtained from personnel familiar with each project by populating a COSYSMO data collection instrument, which included values for all COSYSMO inputs as well as actual effort expended on the systems engineering project. In addition to collecting data on the inputs for the COSYSMO model, the instrument also asked responders about the amount of reuse in the system being reported. At the time, it was not intended for COSYSMO to estimate reuse, but the idea was that it may in the future and the data collection opportunity should attempt to obtain as much information as possible. However, from a modeling perspective, "reuse" was thought of in a much more limited capacity. In the data collection instrument, responders were asked to report the percentage of the value for each size driver that was reused. For example, if a responder reported the project had 100 nominal requirements and that 20 of those 100 requirements where reused, they would report 20% of the nominal requirements were reused. Out of the 42 projects in the data set, 54% reported some amount of reuse in one of the four size drivers and 13% reported reuse in all of the size drivers. With over half of all the projects in the data set reporting some amount of reuse, the potential for COSYSMO to overestimate the effort for projects with reuse became even more apparent.

A significant limitation with this approach of reporting the percentage reused is the lack of consistent definitions and the inability to account for varying levels of reuse. As a result, the various percentages of reuse reported in the data set did not always correlate with expected decreases in total effort. In other words, a single reuse category ("reuse" vs. "no reuse") inconsistently accounted for reuse. Therefore, a single reuse category did not appear to be adequate at accounting for systems engineering reuse.

When COSYSMO was published in 2005, the model was capable (at a minimum) of estimating the effort of a systems engineering project within 30% of the actual, 50% of the time, or PRED(.30) = 50%. To test the hypothesis that a single reuse category is inadequate, an experimental version of the COSYSMO 2.0 operational equation was created. Instead of parameters for six reuse categories, the modified model had a parameter for only one reuse category. Weights for this single reuse category were varied parametrically (from 0.0 to 1.5) and the estimation power of the experimental model was compared with the estimation power of the COSYSMO model over the same set of projects. Across this range of weights, the modified model consistently performed worse than COSYSMO without reuse. The inability of the modified model to improve the estimation power of COSYSMO supports the conclusion that a single reuse category is inadequate. This result also demonstrates the need for multiple reuse categories as well as consistent definitions of the categories.

Appendix F: List of Industry and Government Participants

The Aerospace Corporation
Peter Hantos
Rosalind Lewis
Darryl Webb
Marilee Wheaton

BAE Systems
Jim Cain
John Davies
Allan McQuarrie
Lori Saleski
Alex Shernoff
Gan Wang
Steve Webb
Gary Wisausky

British Petroleum (BP)
Roger Humphreville

Boeing
Rick Cline
Elizabeth O'Donnell
Miles Nesman

Lockheed Martin
Jeff Allen
John Gaffney
Gary Hafen
Garry Roedler
Howard Schimmoller
Kevin Woodward

Master Systems
Stan Rifkin

Massachussetts Institute of Technology
Akshat Mathur
Ricardo Valerdi

Naval Postgraduate School
Ray Madachy

Northrop Grumman
Rick Selby

Rolls-Royce
Shawn Collins

SAIC
Ali Nikolai

Softstar
Dan Ligett

Tecolote
Mike Ross

TI Metricas
Mauricio Aguiar

University of Southern California
Barry Boehm
JoAnn Lane